THE GOLF BIOMECHANIC'S MANUAL

WHOLE IN ONE
GOLF CONDITIONING
3rd Edition

BY PAUL CHEK

ILLUSTRATIONS
CHARLIE ALIGAEN

A C.H.E.K Institute Publication

The Golf Biomechanic's Manual © Paul Chek, 1999, 2001, 2009. All rights reserved.

ISBN-13: 978-1-5387-003-7

Chek, Paul W.
Illustrator: Charlie Aligaen
Editors: Janet Alexander, Cara Burke, Penthea Crozier
      Kathy Jerrit
      Holli Clepper
      Julian Tellames
      Christina Walsh
Cover and Layout: Joling Lee
Proof Reading: Summer McStravick
Technical Advisor: Robert Mottram
Models: Cara Burke, John O'Brien, Robert Yang, Holli Clepper, Franz Sniderman, Raquel Robles, Charlie Aligaen, Andy Ashton
Cover Model: Abner Nevarez

Printed in the United States of America
1st Edition: March 1999
2nd Edition: July 2001
3rd Edition: July 2009

For information contact:
C.H.E.K Institute
Ph: 1.800.552.8789 or +1.760.477.2620
Fax: +1.760.477.2630
E-mail: info@chekinstitute.com
Web: www.chekinstitute.com

No portion of this book may be used, reproduced or transmitted in any form or by any means, electronic or mechanical, including fax, photocopy, recording or any information storage and retrieval system by anyone but the purchaser for his or her own personal use. This book may not be reproduced in any form without the written permission of the publisher, except by a reviewer who wishes to quote brief passages in connection with a review written for inclusion in a magazine or newspaper and has written approval prior to publishing.

Primal Pattern® is a registered trademark of Apriori Anatomikos, Inc, and is used with permission.

CONTENTS

	Acknowledgements	v
	Foreword	vii
	Introduction to the Third Edition	viii
Chapter 1	**Why Condition for Golf?**	**1**
Chapter 2	**Flexibility: A Balancing Act**	**21**
Chapter 3	**Stretching**	**45**
Chapter 4	**How to Warm-Up for Golf**	**77**
Chapter 5	**Functional Exercise**	**91**
	Neuromuscular Isolation – Phase I	98
	Neuromuscular Integration – Phase II	119
	Dynamic Stability – Phase III	130
Chapter 6	**Strength Training**	**149**
	Golf Strength – Phase IV	152
	Golf Strength – Phase V	176
Chapter 7	**Power Training**	**189**
	Golf Power – Phase VI	190
	Golf Power – Phase VII	204
Chapter 8	**Now That You're Hooked!**	**213**
	Resources	219
	References	223
	Index	227

Warning – Disclaimer

The workouts and other health-related activites described in this book were developed by the author and are to be used as an adjunct to improved strengthening, conditioning, health and fitness. These programs may not be appropriate for everyone. All individuals, especially those who suffer from any disease or are recovering from any injury, should consult their physician regarding the advisability of undertaking any of the activities suggested in these programs. The author has been painstaking in his research. However, he is neither responsible, nor liable for any harm or injury resulting from this program, or the use of the exercises or exercise devices described herein.

ACKNOWLEDGEMENTS

No one will ever know how much work goes into writing a book until they actually write one. When you stand in the book store and browse through the many books on the shelves, it is so easy to assume that the author, whose name appears on the front of the book, did it all. I can now attest that this is far from the truth. The book you are now reading is only an expression of my years of clinical experience, constant learning and creativity; an expression that would not be possible without the following very supportive people:

My wife, Penny Crozier, spent countless hours laying out the book, handling numerous computer challenges, studying book publishing and, even more challenging, making my often broken and sometimes illedgable thoughts communicable on paper. It is said that behind every good man is a better woman. Well, my wife will tell you that in front of every good man is a better woman dragging him along, and I will not argue with her. Thank you, babe, for your never ending support and seemingly unending wealth of knowledge. If you ever run for president, I will vote for you.

Cara Burke, telling me that I should attend English classes, and how she must study hieroglyphics to understand my technical notes. Her patience is incredible, as are her editing skills. Janet Alexander, Staff Instructor and Master C.H.E.K Practitioner, for her incredible eye. I think she worked for the secret service as a translator and decoder. No one really knows, but she doesn't miss anything. I commend you Janet for your incredible desire to learn and to help others; you are an angel to many. Shane McDermott, C.H.E.K Practitioner Level 4 and golfer. Shane, thank you for taking the time to carefully read the manuscript and give much valuable feedback that led to the writing of what is now Chapter 1.

Randy Slomovitz, Teaching Pro. Thanks Randy for your often *painfully critical eye!* Thank you for providing me with many useful references and concepts, which aided me greatly in translating my biomechanical and neuromechanical thoughts into concepts that will hopefully change the way teaching professionals view conditioning. Al Vermeil, Conditioning Coach of the Chicago Bulls. Al's many years of experience conditioning golfers made his feedback very valuable. He is truly a leader willing to both share and learn. Thank you for you many years of support Al! Kathy Jerrit, C.H.E.K Practitioner Level 2. Kathy, thanks for being there on short notice when we needed you. Your mastery of the English language and patience with my version of it are appreciated. Additionally, thanks for your feedback regarding technicality and bringing me back to earth with regard to what fitness professionals were likely to understand and apply!

Bob Cisco, PGA Teaching Pro. Bob, thanks for your insight on the game of golf and your appreciation for conditioning and its contribution to the golfer. Your feedback was, and continues to be, very helpful. Jeff Blanchard, D.C., Serious Golfer! Thanks for your useful feedback with regard to the golfer's perspective as well as a chiropractor's insight. Your input provides another dimension to the feedback loop that was helpful. Thanks to Randey Grossman, C.H.E.K Practitioner Level 2, Jordan Grossman, C.H.E.K Practitioner Level 2, and Steve Grossman for over 300 edit corrections Your attention to detail is appreciated, all of you. Thanks also to Dave Dinsmore DDS for his supportive review and constructive input.

VERY SPECIAL THANKS TO Charlie Aligaen, C.H.E.K Level 1 Practitioner and C.H.E.K Institute Artist. As you read this manuscript, you will see close to 200 drawings. They all came from the very talented pen of Charlie, often at Jackie Chan speed! I cannot fully express how thankful I am to Charlie for working countless hours, often every day of the week, to draw, redraw, refine, and even design the many drawings in this book.

Thanks to Tony and Martin, amateur golfers and the barbers at Encinitas Barber Shop. It was they who said "Paul, there is way too much information in here for us man; we just want to play better golf. You don't really expect the average golfer to read this, do you?" It was the feedback from Tony and Martin that made me realize that the book I had written for the public, the book I thought was a simplified version, was a long way from being that. On that day, the decision was made to rewrite the book for the public. That will have to wait until my wife has had a good vacation!

Finally, I'm sure I have missed many people gave me their comments, suggestions and support; I can only say that I am very grateful. Please excuse me for not thanking you all by name.

Sincerely,

Paul Chek, HHP

Editor's note: The first and second edititon contained all illustrations as drawings. In this edition some of the illustrations are replaced by photographs.

FOREWORD

More and more elite coaches and instructors in the game of golf are turning to "cutting edge" technologies to improve the performance of their players. The day of the "Olympic golfer" has arrived, as players become truly better as athletes and golf as a game truly demands athletic endeavor.

As a PGA tour instructor and performance advisor to the pros, I not only have worked with some of the best players in the game of golf today, but see golfers of all levels on a day-to-day basis, all who want to play their best. These golfers all want to make it to the next level in their game.

Yet despite the new technological advances in golf equipment and ball design, there has not been a significant improvement in lowering the scores of these golfers. As a matter of fact, there has not been an improvement of one stroke!

Something is obviously missing in this picture and there is a missing link in the game that goes unhandled and undetected. That "missing link" in my viewpoint is two-fold: firstly, the inability to stay focused and secondly, the lack of physical fitness to play to your potential.

I've had occasion to work with some of the best physical conditioning specialists in the golf industry today with my players who want the edge to stay ahead, and I can say without reservation that Paul Chek, his work and contribution in this field in golf, is right on the cutting edge. I consider myself fortunate to be associated with him; I am a client as well, with my own personalized program that Paul has designed for me. I have made significant improvement and progress in my own game as a professional player. My strength and flexibility have improved dramatically and I feel better physically than I have in years. I'm looking forward to achieving more of my goals as a player personally.

Paul Chek's book, *The Golf Biomechanic's Manual,* will take your game to the next level of golf performance. If you are looking for improved performance and better scores, I strongly recommend that you read and follow this unique program that Paul has devised, that can be tailor-made to your game.

Wishing you all the success in your game of golf.

Bob Cisco
PGA Tour Instructor
Author: *The Ultimate Game of Golf*

INTRODUCTION to the THIRD EDITION

Dear Players, Coaches and Therapists,

Thank you for helping make this book a great success! Since I published *The Golf Biomechanic's Manual* in 1999, I have lectured on golf conditioning at the European Teaching & Coaching conference in Munich Germany, and The Swedish National Golf Association's annual meeting and most recently The World Junior Golf conference in Copenhagen, Denmark.

My approach to conditioning the golf athlete has been well received worldwide. Thank you all for the many calls, letters and emails outlining your successes with the methods within this important book. Thank you those of you that chose to do so for becoming trained to teach these methods professionally.

As many of you know from attending my many workshops around the world, the nutritional components to golf success must NEVER be overlooked. To keep this book manageable to the reader, I have left these very important aspects of nutrition out; yet they are readily available in the following resources available from the C.H.E.K Institute:

- *How to Eat, Move, and Be Healthy!*
- *You Are What You Eat* (audio CD and workbook)
- *The Last 4 Doctors You'll Ever Need – How to Get Healthy Now!*
  (multi-media e-book available from www.ppssuccess.com)

I feel that it is very important that we all do our best to live our dreams fully, while at the same time feeding ourselves with the quality of food that represents our highest commitment to self, planet and game. In this effort, we become "whole" when we are whole-beings, together we can become whole in, and as one!

Love and Chi,
Paul Chek, HHP
Vista, CA January 2009

CHAPTER 1

WHY CONDITION FOR GOLF?

Every golfer wants to play better golf. This desire is a comon thread running from the professional golfer touring on the PGA circuit to the amateur beginner. For many, golf is a chance to relax, relieve stress, do business and exercise. But the wish to lower one's personal handicap and improve one's score is present in even the most recreational golfer. The most common method used to achieve this goal is a combination of professional lessons and more diligent practice. Although this approach seems logical, it is the very reason many golfers end up injured and rarely reach their potential. Why? Simply because few golfers associate the need for improved physical conditioning with their quest for improved performance.

Amateur golfers achieve approximately 90% of their peak muscle activity when driving a golf ball. This is the same lifting intensity as picking up a weight that can only be lifted four times before total fatigue. Yet golfers fail to consider that they strike the ball an average of 30 to 40 times a game with comparable intensity!

The average golfer tends to take up the game at an age when he or she is no longer racing around a sports field, nor actively participating in other energetic, competitive or physically demanding sports. Golf is generally viewed as a game of technical skill rather than an athletic event, requiring less exertion than most other sports. Unfortunately, this common misperception all too often results in injury and/or premature performance plateaus. The reason is very simple: **golf *is* a highly athletic event**! To put this in perspective, consider that the head of a golf club can travel over 100 miles per hour, an effort comparable to pitching a baseball. Or the fact that amateur golfers achieve approximately 90% of their peak muscle activity when driving a golf ball. This is the same lifting intensity as picking up a weight that can only be lifted four times before total fatigue. Yet golfers fail to consider that they strike the ball an average of 30 to 40 times a game with comparable intensity! This level of exertion and muscle activation equates golf with such sports as football, hockey and martial arts. The difference is that other athletes outside of golf include conditioning as an integral part of their preparation for such physical demands.

CHAPTER ONE

Without a comprehensive understanding of the science of strength and conditioning, the golfer who attempts conditioning to enhance performance is likely to seek help from a bookstore or a local personal trainer. Having followed the advice from either source, many golfers find an improvement in their ability to walk from hole to hole, yet are dismayed when there is no improvement in their golf performance. The inevitable downfall is abandonment of strength and conditioning to improve performance.

Without question, there are dozens of books aimed at those seeking better physical conditioning. Unfortunately, most of those books are based on body-building principles, where the primary goal is larger muscle growth. Unlike golf, body-building does not include a functional component; success in body-building is not dependent upon precision timing, control, accuracy or skill. The sole intent of body-building is to become very big and score well with the judges. It has led to a huge industry catering to these demands. Numerous machines and techniques have been developed for isolating muscles and stimulating them to grow – often disregarding the functional aspects of training.

When golfers train using exercise programs based upon body-building principles, sedation of the nervous system's ability to organize and synchronize complex multi-joint movements is inevitable. This result is the complete opposite of what a golfer needs to improve function!

> ***Golfers must consider themselves athletes and train using programs scientifically designed to improve integration and synchronization of the whole body.***

A WORD ON INJURY AND PAIN

When an athlete is injured, careful consideration must be given to the cause of the injury. Often patients are looking for the "quick fix." Doctors and therapists often focus on removing the pain rather than addressing the root cause of the problem. However, when the removal of pain is perceived as the cure, problems are sure to ensue. Injuries recur and many athletic careers end after what should have only been a temporary setback. Recurring pain in the back, shoulder, knee, wrist or elbow is much too common among golfers:

- At any given time, as many as 30% of all professional golfers are playing injured. [1]

- 53% of male and 45% of female golfers suffer from back pain. [2]

- Those who play golf and participate in another sport are 40% more likely to develop back pain than those who just play golf. [3]

By following a carefully designed exercise program that conditions the golfer specifically for the game of golf, the risk of injury may be reduced. Additionally, a golfer already suffering from an injury will have much greater success returning to the game if his program addresses the underlying cause of the injury.

WHY CONDITION FOR GOLF?

IS TECHNOLOGY LOWERING GOLF SCORES?

Have you noticed that the scores achieved by the world's top golf pros have hardly changed in the past fifty years? Granted, many golf courses have been improved or changed during this time, but a quick look through the 1995 *Golfers Handbook*[4] and the 2009 *Golfers Encylopedia*[5] reveals the following:

- The US Masters at Augusta was won in 1942 by Bryon Nelson with a score of 280. In 2008, sixty-six years later, Trevor Immelman won with the same score of 280.

- The British Open at the Old Course at St. Andrews was won in 1960 by Kel Nagle with a score of 278. The 2005 British Open, also held at St. Andrews was won by Tiger Woods with a score of 274. Remember this was the year Woods came back to winning form after adjusting his swing. The second place score in both 1960 and 2005? 279 (Arnold Palmer in 1960 and Colin Montgomerie in 2005).

- In 1966, the Doral Open was won by Phil Rogers with a score of 278. Forty-one years later in 2007, Tiger Woods won the WGC-CA Championship at the same course. His score? 278.

Things are no different with amateur golfers either. According to author Dr. Bob Rotella:

> *Fifteen years ago the average American male golfer's handicap was 16.2. The average female golfer's handicap was 29. TODAY, the average American male golfer's handicap is 16.2, and the average female golfer's handicap is 29!*
>
> Dr. Bob Rotella, *The Golf of Your Dreams*[6]

A very interesting conclusion can be drawn from these facts:

Golfers haven't improved despite technological advances!

A baby elephant is conditioned by its captors to believe that it cannot escape when tied to a small stick in the ground. As the elephant grows up, its captors can keep it tethered to the same small stick, even though the fully grown animal can easily pull a good-size tree out of the ground! Golfers, too, have always believed that to reach their potential, practice and good clubs were all that was required. They have been tied to the belief that high-tech equipment and dedicated practice is all it takes to play the best golf.

CHAPTER ONE

There is no doubt that golf is a game of "action and reaction." The flight and destination of the ball is dependent upon five factors that any good golf professional will point out.

> **Factors that Determine the Flight and Destination of a Ball**
> 1. Clubface Alignment
> 2. Swing Path
> 3. Angle of Attack / Impact
> 4. Speed
> 5. Sweet Spot

What golf pros generally do not know and therefore don't tell you are the physical prerequisites that accurately and consistently enable the player to meet the above five factors. No matter how technologically advanced the equipment, it cannot endow the golfer with a physical capacity that he or she does not possess; even the best clubs don't play the game for you. Unfortunately this is the mental rut that every golfer must avoid.

Consider that a rut is just a grave with both ends knocked out.

Just as the grown elephant can release itself from captivity when it realizes the weakness of its restraint, so too can the golfer break away from previously learned behavior. Golfers need to realize that golf is an athletic game and that they are athletes. Until golfers adopt an athletic attitude and condition for their sport, they will continue to suffer stagnation as demonstrated by the virtually non-existent improvement in golf scores over the past few decades.

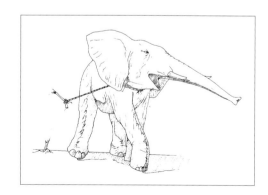

UNDERSTANDING THE WHOLE IN ONE CONCEPT

Whole in One Golf Conditioning is based upon the principles of functional exercise that are designed to restore balance, length, strength and coordination of sports specific movement patterns – in this case, golf. These same principles are successfully used by many of the world's greatest athletes, simply by adapting the concepts to relate to the particular sport in question. Today, the term "functional exercise," like the term "fat free," has been abused by those hoping to attract golfers and other athletes into their machine-based programs. The problem with this is that machines are designed to isolate muscle function. Unfortunately, the brain that controls these muscles does not think in terms of individual or isolated muscles. Rather, the brain recruits groups of muscles in uniquely programmed sequences. Any effective exercise program that is designed to improve function in an athlete, specifically the golfers, must therefore be designed to ***integrate the whole body***. Doing this requires consideration of five key components:

> **Factors in an Exercise Program that Improve Function**
> 1. Flexibility
> 2. Maintenance of Center of Gravity
> 3. Generalized Motor Program Development
> 4. Selection of Open vs. Closed Chain Exercises
> 5. Promotion of Good Posture

Before implementing an exercise program for a golfer, the exercises selected and the program design strategy should be evaluated, considering each of the five components above, to ensure that function will be improved.

1. FLEXIBILITY

To determine which stretches and stretching techniques are most likely to benefit the client, the required ranges of motion (ROM) of specific joints must be known. The specific requirements of the game must be compared against the client's own joint ROM so that the exact stretches needed can be chosen. Use of general stretching prescriptions often results in excessive range of motion in *hypermobile* joints and continued restriction at *hypomobile* joints!

2. MAINTENANCE OF THE CENTER OF GRAVITY (CG)

Golf is played, like most sports, under the influence of gravity in a three-dimensional, unstable environment. If the exercises in any program do not contribute to, or directly enhance the golfer's ability to maintain his or her center of gravity over his or her base of support (feet), the functional carry-over is likely to be minimal. For example, studies have shown that an athlete's ability to squat (a functional movement) is not significantly enhanced by training using the leg press and knee extension exercises. Therefore, these exercises would not enhance function in sports performance.

Golf is a highly technical sport that not only requires maintaining the golfer's CG over his or her base of support at all times; it is a sport of repetitive explosive actions. Unlike shotputting or throwing a discus or javelin, the explosive efforts in golf are directed at a small hole in the ground as far as 550+ yards away. Inability to maintain CG during the golf swing will certainly reduce the chances of maintaining an optimal swing axis or swing path. In such cases, meeting the requirements of optimal ball flight seems unlikely.

Activation of Stabilizers

The body, which can be likened to an inverted pendulum, is very unstable. Although muscle groups such as the adductors, rotator cuff, hip rotators, deep abdominals and deep cervical flexors are known for their roles as postural and stabilizer muscles, nearly all the muscles throughout the body act as stabilizers at one time or another. Of critical importance, with regard to maintaining one's CG over one's base of support, is performing exercises that require the body to stabilize not only what appear to be the working joints, but all joints associated with any given movement.

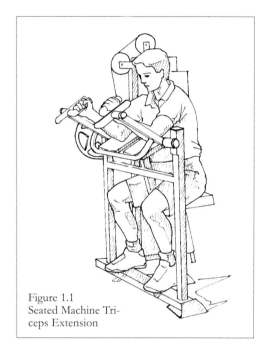

Figure 1.1
Seated Machine Triceps Extension

To exemplify this point, consider the triceps. In the golf swing, the triceps work in concert with nearly every other muscle in the body. Therefore, isolated strengthening in an environment distinctly different from golf will not be very beneficial. The triceps extension exercise as performed on an exercise machine (Figure 1.1) minimizes the requirement for stabilization of the arm by the rotator cuff because the machine stabilizes the load, requiring only force generation in the sagittal (front-to-back) plane.

3. GENERALIZED MOTOR PROGRAM DEVELOPMENT

Generalized motor programs are thought to exist for all movements which display the same relative timing.[7,8] For example, a squat and a jump have the same relative timing, although they are predominantly different in speed and amplitude of movement. This is why the squat exercise improves jump performance. If an exercise is to be functional it should resemble, as closely as possible, one, several or all aspects of the relative timing sequence and movement used to create the full golf swing.

Exceptions to this rule would include specific corrective exercises for the purpose of improving neuromuscular function. If such exercises are being used, then there should be a systematic progression toward exercises with a high degree of similarity in relative timing and movement structure.

WHY CONDITION FOR GOLF?

> **GLOSSARY**
>
> **Open Chain** exercises are defined by the body's ability to overcome resistance, such as during the Lat. Pull Down exercise. As the force generated overcomes the resistance, the handle bar moves towards the body (Figure 1.2B).
>
> **Closed Chain** exercises are defined by the fact that the body cannot overcome the resistance and therefore the body moves away from or toward the resistance. For example, a chin-up or free bar squat (Figure 1.2A).

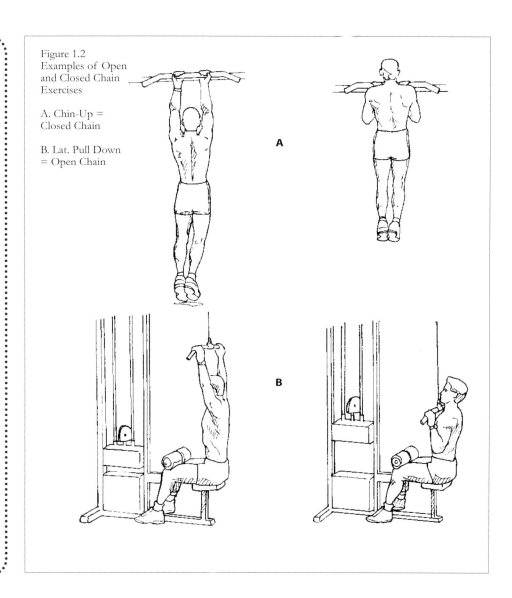

Figure 1.2
Examples of Open and Closed Chain Exercises

A. Chin-Up = Closed Chain

B. Lat. Pull Down = Open Chain

4. SELECTION OF OPEN VS. CLOSED CHAIN EXERCISES

In his exceptional book *Kinesiology Of The Human Body*, Steindler defined the terms **OPEN CHAIN** and **CLOSED CHAIN**.[9] In golf, the upper extremities act in an open chain environment, while the legs work in a closed chain environment. Because the recruitment patterns and joint movements used in an open chain are virtually 180° out of phase with those of the closed chain, selection of exercises requiring the right form of resistance and recruitment pattern will be paramount to success.

CHAPTER ONE

5. POSTURE AND THE GOLFING ATHLETE

What is posture? The word posture is commonly used among golf pros, yet interestingly, the books written by top golfers and teachers are full of players with poor posture. Watch a tournament on TV and it will only take a few seconds to find a top player with forward-head carriage, increased thoracic kyphosis (increased curvature of the upper back), and a flattened or lordotic lumbar curvature (decreased or increased curvature of the lower back). Why then, do you suppose everyone is so mentally conscious of posture, yet not physically? The conditioning specialist is all too often just like the parent reprimanding the child, "Stand up straight boy!"

Posture, like the size of your nose or the shape of your ear, is not something you can just think about and expect to change. Improving posture is a skill that should be the foundation of every rehabilitation and conditioning specialist's approach to exercise and improving function. Fifty years ago, posture was on the forefront of osteopathic, chiropractic, physical therapy and orthopedic studies.[10] The study of posture has been a lifelong pursuit for people like C. L. Lowman,[11] Moshe Feldenkrais, Karel Lewit and F.M. Alexander. It is evident that Ben Hogan was very cognizant of posture when he wrote his 1957 book titled *Five Lessons – The Modern Fundamentals of Golf.* [12]

How Do We Define "Good Posture"?

Before we can begin to restore posture, we must have a working concept of what posture is.

DEFINITIONS OF POSTURE

- *The position of the limbs or carriage of the body as a whole.*
  – Stedman's Medical Electronic Dictionary
- *The position from which movement begins and ends.*

The latter definition is particularly useful when you consider that one's posture is a physical representation of one's psycho-neuro-musculoskeletal set; the physical representation of the organization of body parts as dictated by interaction of the mind (particularly limbic/emotional), nervous system and musculoskeletal system. If you begin and end movement in an aberrant position, the chances of accelerating joint wear are increased.

STATIC POSTURE should not be confused with dynamic posture.

Stated simply, this is the difference between posture when stationary and when moving. A good example of the difference between static and dynamic postures, or task-specific postural expression, is commonly found in professional dancers. Prior to a performance you will often see them sitting in a slouched posture while getting their make-up applied, having a snack, or talking to friends. The instant they walk onto the dance floor, it is as if someone else now inhabits their bodies; their postures suddenly improve to near perfect standards, particularly when in motion.

> **GLOSSARY**
>
> **Static Posture** is the position of the body at rest, sitting, standing or lying.
>
> **Dynamic Posture** is the maintenance of the instantaneous axis of rotation of any/all working joints in any spatial or temporal relationship. (Regardless of the position you are in or how fast you are moving, your joints should always maintain optimal working relationships.)

This is an important point for golfers because sitting or standing in good postural alignment does not guarantee maintenance of the same postural parameters the instant they address a golf ball and perform a full swing. The postural attitude in any given situation is part of the engram, or motor memory, of past experiences of the same task. It becomes evident from this discussion of static and dynamic posture that achieving a healthy postural awareness from both perspectives is necessary.

What Does Good Posture Look Like?

Both the static and dynamic posture of the golfer should be evaluated. Static, standing posture should be examined from two angles; the sagittal plane (viewed from the side) and the frontal plane (viewed from the front or back). Additionally, the address posture should be examined. The dynamic posture during the golf swing is also important, as this may differ dramatically from the static posture.

CHAPTER ONE

Static Posture

From a sagittal plane view (Figure 1.3):

- Imagine a plumb line hanging about 1 cm. anterior to the lateral malleolus (anklebone). We begin at the foot and ankle because they are anchored to the ground, making them a reliable point of reference.
- The plumb line should hang:
- Slightly anterior to a midline through the knee.
- Approximately in line with the greater trochanter of the femur.
- Midway between the back and abdomen as well as the front and back of the chest.
- Through the shoulder joint (providing the arms are hanging in the middle one third of the body).
- Through the lobe of the ear.[13]
- When normal head carriage exists, the sternocleidomastoid muscle will lie at an angle of 45 - 60° relative to vertical.[14]

When viewed from the front or back (Figure 1.4):

- The body should be well aligned, resembling three stacked inverted pyramids.[14,15]
- The pelvis, shoulders and head should sit level with a straight spine.

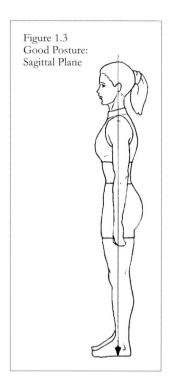

Figure 1.3
Good Posture:
Sagittal Plane

If there are lateral curvatures in the spine (scoliosis), the golfer should bend forward. Look to see if the spine straightens when bent forward, which would indicate a functional scoliosis. If a functional scoliosis is identified, the golfer should have a full length/tension assessment performed and regularly stretch as determined appropriate. (See Chapter 3.) The golfer is then reevaluated three to four weeks later. If the lateral curvatures persist or there is back pain, an orthopedic professional should be consulted.

If during the forward bend test the curvature remains and the ribs appear elevated on one side, a structural scoliosis is suspected, in which case you will need guidance from a proficient orthopedic physical therapist.

In the process of conditioning the golfing athlete, the conditioning specialist or therapist must always seek first to correct static misalignment because of the commonly associated length/tension imbalances that will undoubtedly disrupt dynamic posture.

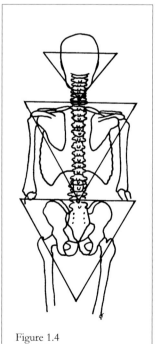

Figure 1.4
Good Posture:
Frontal Plane

The Address Posture

The address posture is described by golf pros as integral to all aspects of the golf swing. Hogan describes initiating the move to address the ball as though you were sitting on a sports stick [16] (Figure 1.5A). He also indicates that the initial bend should come from the knees, keeping them supple, yet at the same time having "live tension" in them. Once in the address position, Hogan indicates that your weight should be more toward the heels than the balls of the feet, "so that if you wanted to you could lift your toes in your shoes" (Figure 1.5B).

> **Note:** The conditioning specialist may have used this technique when cueing a client to lift the toes prior to squatting, for the purpose of having him or her squat with a more upright posture. [17]

While at address, or what Cisco [18] refers to as the set-up position, it is very important not to excessively protract, or round, the shoulders. Protraction of the shoulders is mechanically coupled with increased thoracic kyphosis and encourages forward head posture; both of these unwanted postural aberrations will restrict spinal rotation. This hypothesis can easily be tested by sitting on a bench with a dowel rod or golf club held across your chest. Sit first with increased kyphosis and forward head posture, then rotate the trunk as far as possible without moving the pelvis (Figure 1.6A). Now, assume an upright posture with natural spinal curvatures and normal head carriage; note that upon rotating your trunk you achieve greater range of motion with less muscular effort (Figure 1.6B).

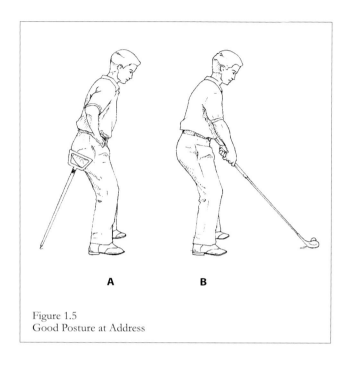

Figure 1.5
Good Posture at Address

Figure 1.6
Testing Relationship Between Spinal Curvatures and Rotation
A. With increased kyphosis
B. With good upright posture

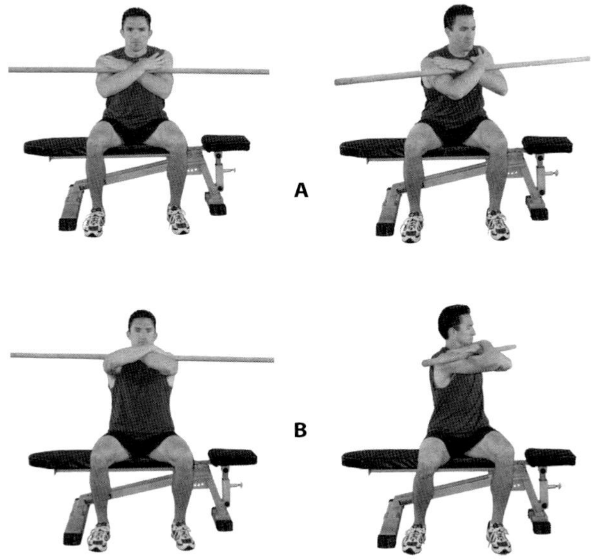

CHAPTER ONE

When in the address posture, a plumb line dropped from the golfer's shoulder should bisect the base of support (Figure 1.7). Care must be taken not to round the upper back when looking down at the ball. The head only needs to tip downward from the upper cervical joints to bring the gaze to the ball.

Interestingly, the posture of the arms as described by Hogan is almost identical to the arm and shoulder position optimal for initiation of a dead lift or power clean.[19] Just as turning the anticubital space of the elbow forward activates the scapular adductors to stabilize the load in the dead lift, this posture also aids the golfer in integrating the arms and trunk as a functional unit, while discouraging unwanted kyphosis (Figure 1.8).

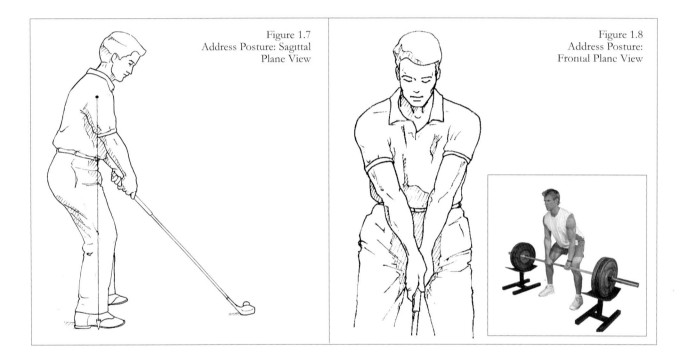

Figure 1.7
Address Posture: Sagittal Plane View

Figure 1.8
Address Posture: Frontal Plane View

GLOSSARY

Protract the Shoulders - to round the shoulders forward.

Thoracic Kyphosis - an increase in the curvature of the section of spine between the lumbar and cervical spines (the mid to upper back).

Forward Head Posture - a forward displacement of the head and neck in the sagittal plane, away from the position of good posture.

WHY CONDITION FOR GOLF?

THE GOLF SWING FROM A DYNAMIC POSTURE PERSPECTIVE

Which muscles and joints are involved in the golf swing? Well, a safe answer would be all of them. With only a basic understanding of kinesiology and biomechanics, anyone studying the full swing would certainly come to the conclusion that it is complex, to say the least. In consideration of the description of "dynamic posture" already given, it is evident that maintenance of the instantaneous axis of rotation or optimal joint function during the full swing is neuromechanically challenging – even under ideal conditions.

To fully understand what is happening in the golfer's body, we must consider what it takes to have a low handicap. A low handicapper must be consistent. To be consistent, you must be able to reproduce a good swing consistently. To appreciate the neuromechanical challenge, let's look at some of the basic kinesiological and biomechanical requirements that must be met.

The Backswing, Downswing and Follow-through

From address, the player initiates movement of the club with the arms, as though to sweep a second ball away (Figure 1.9). As the arms rise, following an imaginary swing-plane running from the top of the shoulders to the ball, the torso, followed by pelvis, completes the sequence that ends when the player's club approximates the "doorframe position" (Figure 1.10).

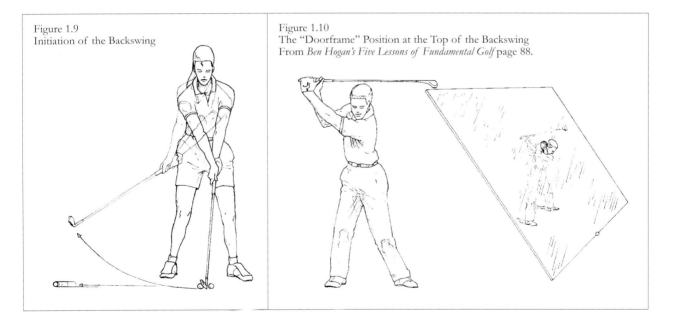

Figure 1.9
Initiation of the Backswing

Figure 1.10
The "Doorframe" Position at the Top of the Backswing
From *Ben Hogan's Five Lessons of Fundamental Golf* page 88.

CHAPTER ONE

What is of importance is that as the backswing takes place, there is transference of about 75% weight to the right leg, yet the leg should not shift laterally (Figure 1.11). As the arms, trunk and pelvis are rotating (often referred to as "coiling"), there should be minimal shift of the axis or rotation (Figure 1.11). An excessive shift is easily identified as excessive lateral movement of the head in relationship to the ball (Figure 1.12). Should there be a lateral shift of the head greater than about half a head width, the chance of achieving optimal swing factors and ball flight decrease exponentially!

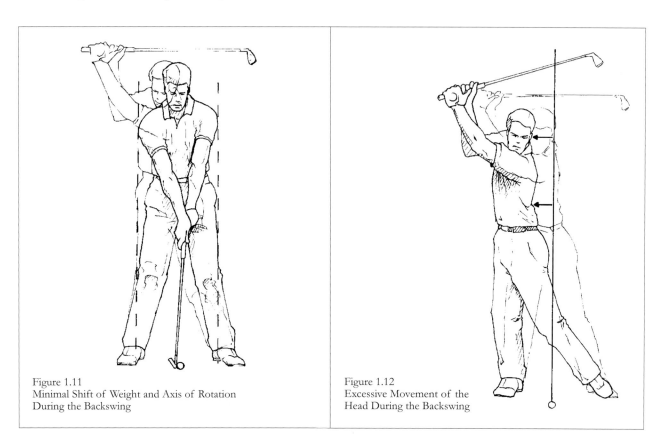

Figure 1.11
Minimal Shift of Weight and Axis of Rotation During the Backswing

Figure 1.12
Excessive Movement of the Head During the Backswing

WHY CONDITION FOR GOLF?

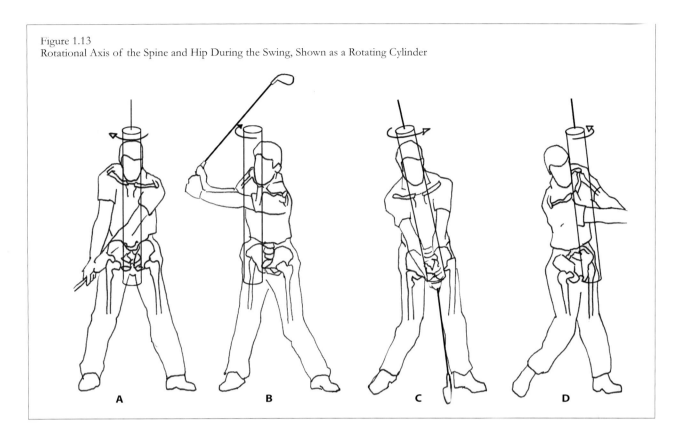

Figure 1.13
Rotational Axis of the Spine and Hip During the Swing, Shown as a Rotating Cylinder

When muscle balance and flexibility are optimal, the player should be able to perform the backswing with a dual rotational axis synkinetically orchestrated between the spine and the right hip (Figure 1.13 A-B). As the coiling action takes place and weight shifts toward the right leg, the rotational axis progresses toward the hip joint. A skilled golfer will almost be able to combine the axes to become one large one, looking somewhat like a rolling cylinder (Figure 1.13 C-D). As this cylinder reaches its end point, it quickly changes direction and the rotational axis progressively shifts to the opposite side during the downswing and throughout the follow-through.

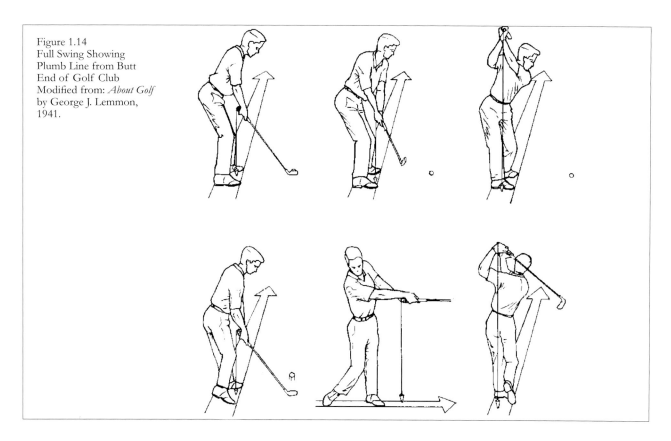

Figure 1.14
Full Swing Showing Plumb Line from Butt End of Golf Club
Modified from: *About Golf* by George J. Lemmon, 1941.

A plumb line dropped from the butt end of the club shaft should fall on a line bisecting the player's feet at any point during a properly executed full swing (Figure 1.14).

Without mentioning the obvious orthopedic requirements, the challenge of keeping optimal visual focus is worth discussing briefly. During the backswing, particularly, the head must remain fairly stable to allow optimal target acquisition. Should there be any restriction of motion in the upper cervical, lower cervical, and even the upper thoracic spine (which behaves like the cervical spine to the level of T3-T5, the third and fifth thoracic vertabrae), one of three things must happen:

- The body must compensate for lost motion under the cervical restriction. This will affect swing consistency.
- If the restriction is below C2, the second cervical vertabrae, there may be compensation at C1/C2 and/or C0/C1. Such compensations may significantly hamper proprioception, as this region of the spine is a contributor to both balance and proprioception.
- Should the body be unable to compensate effectively, the eyes will be momentarily pulled from the ball, disrupting target acquisition and body control.

During this entire process, the five factors governing ball flight must be optimally controlled or the golfer will get to go on a "walk about," or even a little swim to keep the ball in play.

WHY CONDITION FOR GOLF?

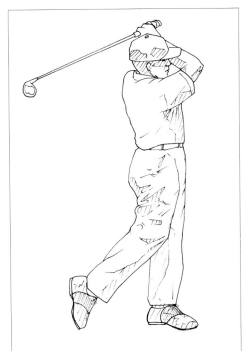

Figure 1.15A
Good Follow-Through Posture - sagital view

Figure 1.15B
Good Follow-Through Posture - frontal view

Once ball compression has taken place, follow-through will be executed. In simplistic terms, follow-through may be thought of as a mirror image of the backswing, although each player has his unique follow-through. Ideally, the follow-through should terminate in a well-balanced, upright posture (Figure 1.15). The reverse "C" follow-through posture is to be avoided because it places excessive sheer forces through the lumbar spine, which has been found to be clinically related to back pain in golfers.

POSTURE, GRAVITY AND GOLF

Now that you have had a brief explanation of the complex processes of the body that are involved with the full golf swing, a short discussion of gravity, posture and spinal function in the golfing athlete is warranted. Golf requires full rotational capacity of nearly every joint involved. Golf is a rotation sport; to reach your golf potential, you must be able to rotate repeatedly efficiently and explosively.

When a player rotates his or her trunk, there is tremendous gravitational load on both the spine and spinal cord. To enhance this point, consider that when supine (lying down on your back), the gravitational stress on the spinal cord is approximately 30 lb/inch,[19] whereas in the standing position the gravitational stress ranges from 190 to 300 lb/inch.[20,21,22] Vertebrae that are aligned to the vertical axis execute normal motion (maintenance of instantaneous axis of rotation), whereas vertebrae whose centers of motion have deviated from the vertical axis execute abnormal motion.[23,24]

With this in mind, consider a known physical principle: force applied to an object imparts to it an acceleration, not only in translation but also in rotation; the object turns around its own center of gravity.[25,26] This is important as rotation may cause the affected vertebrae to displace from the normal position; chiropractors and osteopaths refer to the misalignment of such vertebrae as subluxation.

Golfing athletes with inadequate postural alignment, muscle imbalance syndromes and associated joint motion restriction will not be able to efficiently rotate. This was proven by Feldenkrais who, in comparing humans and animals,[27,28] calculated mathematically the moment of inertia of the body around the vertical axis.

CHAPTER ONE

Figure 1.16
Beary Interesting Posture!

Feldenkrais noted that when animals adopt a bipedal stance, the head leans forward and is balanced by the pelvis protruding backward. The result is that the moment of inertia around the vertical axis is four to five times greater than human-like vertical alignment. Animals (like humans with very poor posture) are very clumsy in their attempt at vertical stance and are almost incapable of rotational movement (Figure 1.16). In conclusion, he states that the body is mechanically most efficient when it is held so that it can turn itself around with the least effort (i.e. the configuration that has the smallest moment of inertia around the vertical axis).

The position in which the pelvis, trunk, and head are aligned vertically and the spinal curvatures are at a minimum is the position in which minimum muscular contraction is necessary to maintain the body from falling over. The least muscular tone is consequently required in that posture.[29] As the golfer deviates from ideal spinal alignment, postural tone increases, as well as energy requirements, causing performance to decrease and the incidence of musculoskeletal pain to increase.

FLEXIBILITY, STABILITY, STRENGTH & POWER

The following are four physical factors that must be addressed in order to be able to provide the four key components necessary to control ball flight (clubface alignment, swing path, angle of attack and speed).

> 1. Muscle Balance and Flexibility
> 2. Static and Dynamic Postural Stability
> 3. Strength
> 4. Power

It is important to address these factors in the correct order. The first step in the Whole in One Golf Conditioning program is to improve flexibility, as this is a catalyst for all subsequent aspects of golf conditioning. Once flexibility is restored to appropriate areas and the musculoskeletal system is balanced, static and dynamic postural stability can be acheived. Stability is important since a stable body creates a solid framework for all movements and activites. Additionally, a stable, well-balanced body is less likely to be injured. When stability is achieved, strength can be built using functional movement patterns that will readily transfer to the game of golf. Finally, the last progression is to develop power – the more power a golfer can transfer from his body through the club to the ball, the farther he will be able to drive the ball. Any attempts to improve golf strength or power without first restoring flexibility and stability will often prove to be less fruitful and more likely lead to injury!

SUMMARY

Too many golf conditioning books and programs break one or more of these principles. The result is either little or no improvement in performance, or worse still, a decrease in performance or actual injury. In contrast, Whole in One Golf Conditioning is based upon the **Flexibility - Stability - Strength - Power** progression, allowing the golfer to condition for the game in a manner that is conducive to optimal golf performance.

To truly improve performance, prevent and treat injury there are a few simple principles that must be followed:

1. All treatment or exercise programs must seek to correct the underlying cause(s) before trying to improve performance.

2. Treatment should only focus on alleviation of pain during the acute phase of the injury. The acute phase generally lasts no longer than three weeks.

3. If restored function and/or improved performance is the goal, a designed exercise program must first restore flexibility and stability to the working joints and balance the musculoskeletal system. Next, strength is restored followed by re-establishing power output.

There are a number of other program design factors that influence the outcome of any exercise program that are beyond the scope of this book. However, the guidelines given here serve to show the fabric upon which the Whole In One Golf Conditioning program is based; they are the principles upon which any program claiming to be functional should be based. It should come as no surprise to the reader that these very principles apply not only to every athlete, but every human being; they are universal.

CHAPTER ONE

HAS YOUR SCORE IMPROVED IN THE LAST TEN YEARS?

Write down your average score now, one year ago, five years ago and ten years ago. If you have been playing for less than ten years, fill in the spaces that relate to you. Apart from the time when you were first learning to play, how much has your score improved? What would you like your score to be? Write that in the space provided.

Your Ideal Score	_____
Score Now	_____
1 Year Ago	_____
5 Years Ago	_____
10 Years Ago	_____

There is no faster way to improve your golf score than improving you, the golfer! No club in the world will match the performance gains achievable through the Whole In One Golf Conditioning program.

CHAPTER 2

FLEXIBILITY: A BALANCING ACT!

Golf flexibility is not something that can be purchased at the pro shop. It is, however, essential in reaching a golfer's full potential. Some people have good flexibility naturally. For those who don't, flexibility is needed and can be developed. Adequate flexibility is especially important for golfers who suffer back, wrist, shoulder, hip and even knee pain during or after golfing. Golf flexibility is a descriptive term for the amount of movement, uninhibited by range of motion restrictions, that a golfer needs to play golf to his or her full potential.

An understanding of the three planes of movement is useful. Without much imagination, it is clear that all three planes are used in golf and the ability to move unrestricted in each is important. If the golfer has problems (including such common faults as posture at address, achieving optimal back swing, swing timing, finding optimal swing plane, follow through, hooks, slices, hitting fat or thin, putting, chipping, pitching or driving straight), then stretching is likely to help.

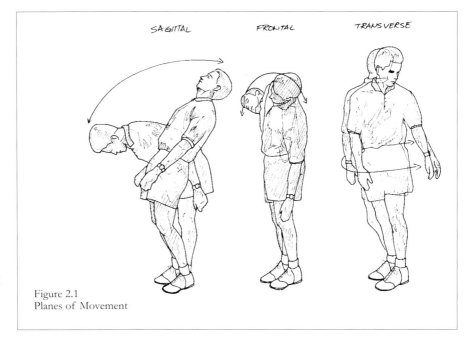

Figure 2.1
Planes of Movement

CHAPTER TWO

HOW STRETCHING HELPS THE GOLFER

Stretching allows the development and maintenance of optimal joint range of motion in the golfer's body. This is important since an effective golf swing depends upon the ability of the golfer's joints to move through the necessary planes of motion. In other words, optimal joint range of motion is a biomechanical prerequisite of the golf swing. The tissues that affect swing mechanics are the muscles, tendons (together the musculotendinous unit), ligaments and joint capsules surrounding each of the joints involved. All these tissues, with the exception of ligaments, have the ability to shorten. This is generally an undesirable event in a golfer. (Ligaments generally are not stretched except in special cases, for example after injury.)

When the musculotendinous unit and/or joint capsules shorten, the body's biomechanics are altered, progressively disrupting swing mechanics. This happens because muscles attach to both sides of a joint and act synergistically. Depending upon the joint, some muscles are responsible for medial rotation, while the opposing muscles produce lateral rotation, as is the case with the shoulder (Figure 2.2). Others produce flexion, countered by those producing extension. When the synergy, or balance, of these muscle groups is altered, the mechanics or movement pathways of the joint are also changed. To the golfer, this means compensation. The more a golfer compensates for altered joint mechanics, the less consistent he or she will be and the more swing faults will haunt his or her game.

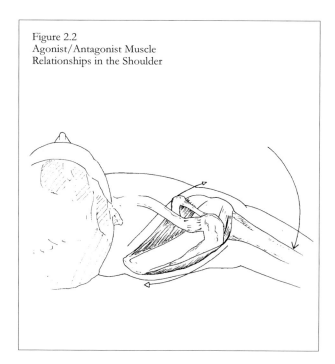

Figure 2.2
Agonist/Antagonist Muscle Relationships in the Shoulder

GLOSSARY

Muscle:
Elastic tissue made of many fibers that act across a joint to cause movement

Tendon:
Connects tissue to bone

Ligament:
Connects bone to bone across a joint

Joint Capsule:
Connective tissue that surrounds a joint

Medial Rotation:
The limb turns inward toward the midline of the body, also called internal rotation

Lateral Rotation:
The limb turns outward away from the midline of the body, also called external rotation

FLEXIBILITY

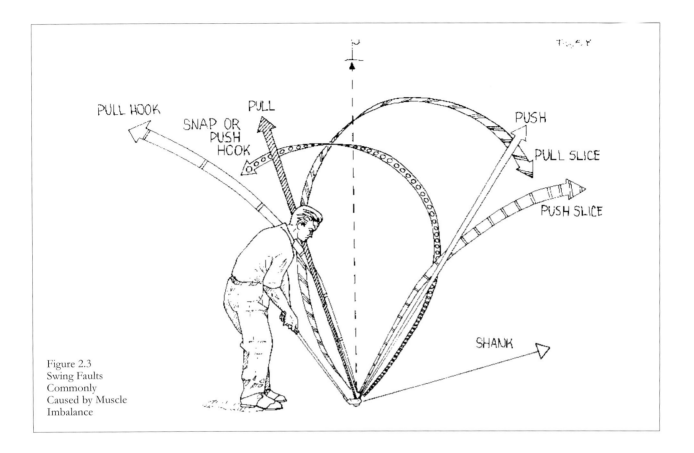

Figure 2.3
Swing Faults
Commonly
Caused by Muscle
Imbalance

CHAPTER TWO

GOLF IS ROTATION

To golf at full potential, the golfer must possess the ability to rotate almost every joint to its functional capacity. If there are movement restrictions in the shoulder girdle, torso, pelvis or hips, there will be compensation somewhere else in the musculoskeletal system. The result of such compensation is most often seen as faults in the golf swing. These compensations may also express themselves in the form of pain and/or injury. To better clarify the point, consider the following examples:

- If the scapulothoracic joint (shoulder blade on rib cage) is limited in motion, then the glenohumeral joint (shoulder joint) will have to compensate (Figure 2.4). Should compensation occur at the shoulder joint, the chances of hooking or slicing the ball increase dramatically.

- A very common problem in golfers is restricted shoulder range of motion. Restricted shoulder motion is often compensated with excessive spinal rotation. This frequently leads to back injury because most golfers already lack rotational flexibility in the spine. When the shoulder is restricted and the spine cannot effectively compensate due to limited rotation, the golfer commonly has problems with excessive head motion or maintaining an optimal swing plane, and frequently hits either fat or thin (Figure 2.5).

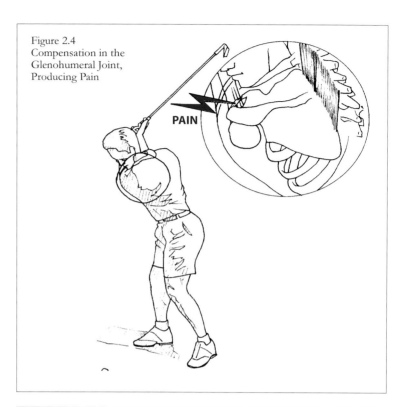

Figure 2.4
Compensation in the Glenohumeral Joint, Producing Pain

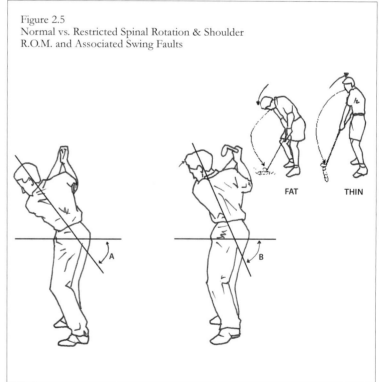

Figure 2.5
Normal vs. Restricted Spinal Rotation & Shoulder R.O.M. and Associated Swing Faults

FLEXIBILITY

- If the hip joints are restricted in either internal or external hip rotation, then golf swing form will be affected and excess rotational demands will be placed on the spine or shoulder joints. When the body does not have the capacity to effectively compensate at either the shoulder or spine, the golfer is forced to overuse the wrists to drive and decelerate the golf club. This is one of the major sources of wrist injury in golfers.

- Another common compensation for decreased range of motion at the hip and back is excessive elevation onto the toes of the left foot during the back swing. This inevitably results in fat shots due to a chopping action on the down swing (Figure 2.6).

The most common by-product of reduced flexibility in the shoulder girdle, torso, pelvis or hips is reduced power. Reduced power means less distance on the drive, something that no golfer wants. By performing a proper golf-specific stretching program, injury can be avoided and performance improved in almost every aspect of the game.

In fact, as a golfer's specific flexibility improves, he is much more likely to benefit from golf lessons. If there is a biomechanical fault underlying the swing fault, chances are slim that long-term resolution will be achieved through lessons. In reality, all a golf pro can do is try to find other ways to compensate! This means that a lot of time is spent **programming the nervous system** with less than optimal movement patterns. At some point, these faulty movement patterns have to be replaced with correct ones. Breaking old habits is much harder than developing new ones.

Figure 2.6
Decreased Range of Motion in the Hip & Back
The golfer on the left has only 20° of rotation in the hip and back. To produce the 40° of movement needed to complete the shot, the golfer on the right may lift up the heel of the left foot.

GLOSSARY

Programming the Nervous System
The body remembers movement patterns, but cannot distinguish between good or poor patterns. When in a situation where a movement has to be reproduced, the nervous system will provide the closest pattern, whether this is optimal or not. It is therefore better to train with good movements from the start, rather than having to re-learn them later.

CHAPTER TWO

WHEN TO STRETCH FOR OPTIMAL GOLF PERFORMANCE!

There are five main categories of stretching the golfer must be familiar with. These categories are described in detail in Chapters 3 and 4.

> 1. Developmental
> 2. Maintenance
> 3. Pre-Event
> 4. Post-Event
> 5. Muscle Energy Mobilization

Understanding the science of flexibility and golf mechanics is quite complex. There are hundreds of stretches available in many books and videos along with many schools of thought regarding the "correct way" to stretch.

Time is another hurdle to deal with when it comes to stretching and exercising. There are so many possible stretches. To perform all of them would require the golfer to give up golf so he or she has time to stretch. Selecting the stretches most likely to improve the golfer's game will prove effective as a time management tool and also produce the best results. This is a four-step process.

> ### 4 STEPS TO SUCCESSFUL STRETCHING
>
> 1. Identify the short tight muscles.
>
> 2. Select stretches to restore optimal golf flexibility and golf mechanics.
>
> 3. Correct flexibility with *Developmental Stretching*.
>
> 4. Once normal range of motion is restored, maintain flexibility with a combination of *Maintenance, Pre-Event* and *Post-Event Stretching*.

IDENTIFYING SHORT TIGHT MUSCLES:
The First Step Toward Optimal Swing Mechanics

Have you ever been frustrated because your body would not produce the shot your mind was visualizing? Chances are that your mind did send a perfectly sequenced command to your muscles, but the sequence got altered along the way. To better understand the breakdown between mind and muscles, some basic concepts of muscle imbalance and how the nervous system activates the muscles need to be explored.

When the body is out of balance, there are some muscles that are shorter than normal and some that are longer than normal. Muscles are composed of fibers that are slow contracting (slow twitch) and fast contracting (fast twitch). Many muscle groups have a dominance of either slow or fast twitch muscle fibers, while a few muscles have an equal proportion of both.

Researchers in the field of musculoskeletal medicine have found that the response to faulty loading differs among muscles. Faulty loading refers to any form of overuse, under-use or trauma to the musculotendinous unit or related joint complex. These researchers have identified two particular categories of muscles of interest to the golfer: tonic and phasic.

Additionally, the nerves feeding the tonic muscles have a lower threshold for stimulation than phasic muscles.[1] This means that once tonic muscles get in the habit of doing the work for phasic muscles, the nerves supplying the tonic muscles tend to rob portions of the messages sent to the phasic muscles from the brain. These nerves, with their lower thresholds, act like low threshold conductors in an electrical circuit (Figure 2.7). In this example, the brain sends out a message for a specific movement. The low resistance tonic back muscles receive greater input than the higher resistance phasic abdominal muscles and produce a distortion of the original message sent by the brain. Instead of a perfect shot, the end result shows up as a hook, slice, thin, fat or some other poor shot.

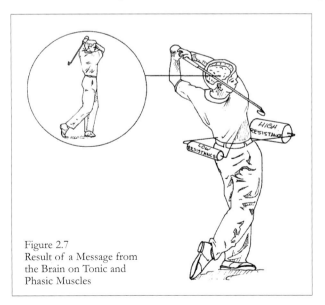

Figure 2.7
Result of a Message from the Brain on Tonic and Phasic Muscles

GLOSSARY

Tonic Muscles react to faulty loading by shortening and tightening. They have a tendency to try to do the work for the opposing or synergistic phasic muscles. An example of tonic muscles would be the lumbar erector (low back) muscles.

Phasic Muscles tend to lengthen and weaken in response to the same stress that shortens tonic muscles. An example of phasic muscles would be the abdominal muscles.

CHAPTER TWO

When this happens, golfers naturally think they must modify their stance, swing plane, grip, or some aspect of the swing pattern to prevent such an event next time they swing the club. Sometimes this works but, in most cases, only for a short time. The next time the golfer tries the same modification, he or she may get a completely different response. This is because the degree of balance between the muscles is constantly changing. Muscle balance is affected by such things as mood, temperature, level of arousal, stress, hormonal balance and stimulants such as sugar and caffeine. Alcohol may also affect muscle balance.

To effectively balance your swing, you must identify your particular pattern of muscle imbalance. Each individual's muscle imbalance presents a unique pattern. This pattern is dictated by such factors as past trauma, work environment, repetitive exposure to sport specific stresses and posture.

> *Muscle balance is affected by such things as mood, temperature, level of arousal, stress, hormonal balance and stimulants such as sugar and caffeine.*

The following series of muscle balance tests will help the golfer identify which muscles must be stretched in order to achieve balance in the golf swing.

KEY FOR THE TEST SYMBOLS

📋 = The clipboard represents the explanation of the test.

✓ = The check mark represents a normal finding to the test.
(Clinically, this is called a negative finding.)

!! = The double exclamation marks represent a positive finding to the test and the need to correct a fault. It also indicates when and which stretches will need to be implemented to correct the findings. (Clinically, this is called a positive finding.)

🚩₁₈ = The flag represents how the test findings are applicable to your golf game.

Neck Side Flexion Test

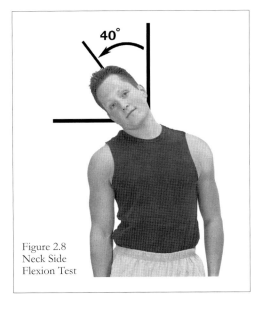

Figure 2.8
Neck Side
Flexion Test

- Standing, or seated in front of a mirror, side bend your neck as though bringing your ear toward your shoulder. Be very careful not to allow your shoulder to tip as you side bend your neck and head.

✓ If you have normal range of motion you will be able to side bend your head between 25 - 40°.

!! If you can side bend to one side noticeably more than the other, you need to stretch the tight side (see Figure 3.2 for the Neck Side Flexor Stretch). If after a couple of weeks of stretching two times daily there is no change, it would be wise to see a C.H.E.K Practitioner. If there is not one available near you, consult a registered physical therapist, chiropractor or related medical practitioner.

Neck Rotation Test

- Start by sitting upright in a chair, maintaining good posture. Be sure to keep your shoulders and back against the chair. Rotate your head to the right and then to the left.

Figure 2.9
Neck Rotation Test

✓ You should be able to rotate at least 70-90° in both directions. Ninety degrees of neck rotation is achieved when you can rotate your head until your chin lines up with the seam of your shirt as it runs over your shoulder. If you are age sixty or more, 70° is the norm.

!! If you can't rotate your head through the suggested range of motion, perform the Neck Rotation Stretch (Figure 3.1) before your game and as part of your developmental stretching program.

▶18 If you are restricted in neck rotation, you most likely have to take your eyes off the ball momentarily during the back swing. This may result in loss of swing plane and/or faulty clubface angle at impact. (For an expanded understanding of the eyes see Chek, Reference 2.)

CHAPTER TWO

Sweetheart Test

The Sweetheart Test is used to determine if the levator scapulae muscles are short. The levator scapulae muscles are those that lift the shoulder into a shrug and will restrict neck rotation if tight. They also assist in maintaining optimal position of the shoulder joint during movement.

Figure 2.10A
Sweetheart Test

📋 To test the length of the levator scapulae muscles, first perform the Neck Rotation Test (Figure 2.9) to assess your range of motion. You then place one arm around a partner, making sure to completely relax the muscles in your arm and shoulder. Rotate your head away from your relaxed arm. Switch arms and repeat the test on the other side (Figure 2.10A).

You can also use a partner to help test the levator scapulae. Sit in good, upright posture. Look to the left and note how far you can rotate your head. Repeat to the right. (Figure 2.10B-a). Ask a partner to lift your elbows until your upper trapezius muscles are relaxed. Now look to the right and left again (Figure 2.10B-b).

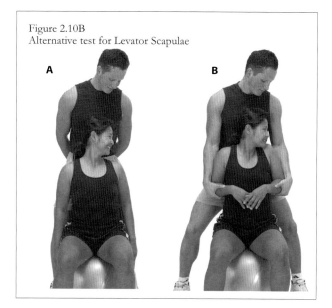

Figure 2.10B
Alternative test for Levator Scapulae

✓ There should be no difference in the amount of head rotation when the arms are supported or unsupported in either of the two tests.

!! If your range of motion improves on either side with the Sweetheart Test, then that levator scapulae muscle needs stretching. (See Figure 3.3 for the corresponding stretch.)

With the partner-assisted test, if your range of motion increases when your elbows are supported and shoulders lifted, this indicates a tight levator scapulae on the opposite side to the way you are looking.

FLEXIBILITY

- A positive Sweetheart Test indicates that the shoulder girdle will not be able to effectively rotate around the spine and rib cage. During the back swing, levator scapulae tightness will restrict left neck rotation, which is necessary to keep your eyes on the ball. A shortened levator scapulae on one or both sides will also cause unwanted compression in the neck with arm movements. This may manifest as tension headache and/or neckache after playing golf.

- To evaluate the effect of a shortened levator scapulae on swing mechanics, simply stretch this muscle immediately before hitting the ball (Figure 3.3) and assess the changes in the swing. Regardless of whether the test improves the swing mechanics, if it changes the swing noticeably in any way, the stretch should be frequently performed before playing. Over time, this will retrain the body and teach it how to play with normal levator scapulae length.

Apley Scratch Test

The Apley Scratch Test (Figure 2.11) is used to assess both internal and external rotation of the shoulders.

- This is a two-part test: First, reach over the shoulder and attempt to touch the top inside corner of the opposite shoulder blade. This is the test for external rotation (Figure 2.11A). Then reach behind the back and attempt to touch the lower part of the opposite shoulder blade. This tests internal rotation.

- The ability to reach behind the neck to touch the top inside corner of the opposite shoulder blade indicates normal external rotation.[5] Normal internal rotation is indicated by the ability to reach behind your back and touch the inferior angle of your opposite shoulder blade (Figure 2.11B).

- The greater the distance between your fingers and the shoulder blade, the tighter the rotator muscles are. If you are restricted in external rotation (Figure 2.11A) your medial or internal rotator muscles of the shoulder are tighter and you will perform stretches in Figure 3.5A+B. If you are restricted in internal rotation (Figure 2.11B) your lateral or external rotator muscles of the shoulder are tight and you will perform the corresponding stretches in Figure 3.6.

- If you have external rotator tightness in the right shoulder, your follow-through will be restricted. External shoulder rotator tightness in your left shoulder will result in restriction of your backswing. Tightness in the left external shoulder rotators will have the greatest impact on the flight of the ball, as compensation is often seen as loss of your swing axis with swing plane alterations.

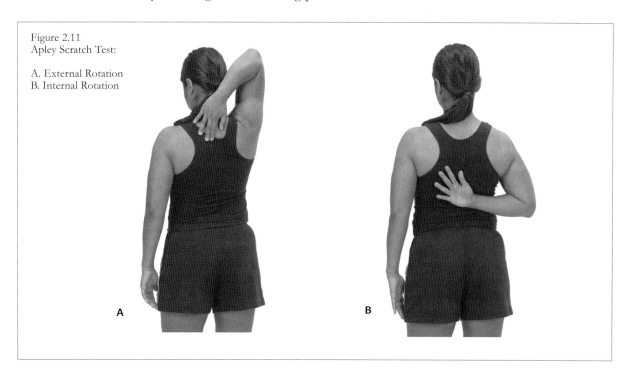

Figure 2.11
Apley Scratch Test:

A. External Rotation
B. Internal Rotation

FLEXIBILITY

Pectoralis Minor and Major Test

☐ To perform the test for tightness of the pectoralis minor and major muscles, lay on your back with your hands behind your head. Allow your arms to drop toward the floor.

✓ Normal pectoral range is when the forearms lay flat on the floor (Figure 2.12B).

!! If both forearms do not lay flat on the floor (Figure 2.12A), your pectoralis minor and major are tight and need to be stretched. It is also possible to have tightness on one side only (Figure 2.12C). Figure 2.13 shows tightness in the left pectoral with the arms to the side. See Figures 3.5B and 3.8 for the corresponding stretches.

▶18 Shortness in the pectoralis minor and major muscles will produce a combination of those faults identified for both the external shoulder rotators and the levator scapulae muscles. Therapists and conditioning professionals should be aware that pec minor shortness may contribute to or exacerbate anterior capsular laxity, secondary impingement syndromes and instability. This often results in pain in the backswing or follow-through positions.

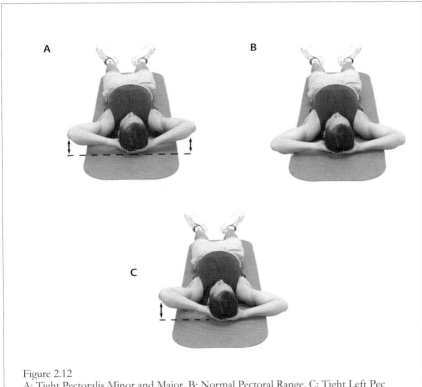

Figure 2.12
A: Tight Pectoralis Minor and Major, B: Normal Pectoral Range, C: Tight Left Pec

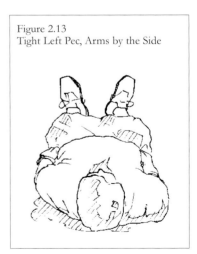

Figure 2.13
Tight Left Pec, Arms by the Side

CHAPTER TWO

Spinal Rotation Test

- To perform the Spinal Rotation Test, lay on your back with your legs in the air, bent at the knees. Slowly lower your legs to one side.

- If you have normal spinal rotation, your legs will lie flat on the floor without the opposite shoulder coming off the ground (Figure 2.14A).

- The greater the distance your bottom leg is from the floor when the opposite shoulder begins to rise, the more restricted your spinal rotation (Figure 2.14B).

- Restricted spinal rotation will result in excessive internal shift and rotation of the hips during both the backswing and follow-through. The shoulder is often overused to compensate for restricted spinal rotation. When the spine can't fully rotate, coil action is limited. Limited coil results in attempting to accelerate the club with the arms and is a common source of golfer's elbow. It can also produce any swing faults with regard to swing plane, clubface angle and maintenance of optimal swing axis. The corrective stretch is described in Figure 3.14.

Figure 2.14A
Normal Spinal Rotation

Figure 2.14B
Restricted Spinal Rotation

FLEXIBILITY

Thomas Test

To perform the Thomas Test at home, find a strong table or firm bed to lie on (Figure 2.15A). Place your body so that you are lying on your back with your legs hanging off the end of the bed or table. Make sure your knee and lower leg can hang freely without any obstructions. Place one hand under your lower back, opposite your belly button. Using the other hand, bring the knee to your chest. Pull your leg back until you feel your spine begin to press down on the hand under your back. At this point, look to see if the other leg has lifted off the table or bed. Check also if the lower leg is hanging straight down toward the floor.

✓ If the thigh of the leg on the table remains flat and the lower leg is perpendicular to the floor, then you have passed the test (Figure 2.15A). When this happens, you can begin a maintenance stretching program. (See Chapter 3.)

!! If the thigh of the leg opposite to the one you have pulled toward your chest has come off the table (Figure 2.15B), you need to perform the Lunge Stretch (Figure 3.15). If the lower leg does not hang straight down toward the floor, you need to perform the Quadriceps Stretch demonstrated in Figures 3.25A, B &C .

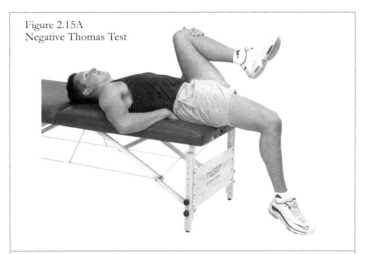

Figure 2.15A
Negative Thomas Test

A positive Thomas Test indicates short hip flexors. Short hip flexors can limit your ability to achieve a full backswing, reducing your ability to turn the trunk. The result is often a loss of distance on your drives. Your follow-through may also be limited, which can impede your ability to execute a straight shot and can limit distance as well.

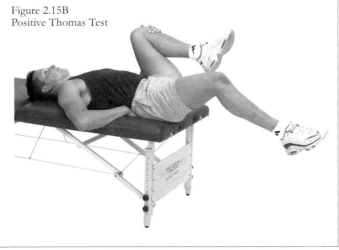

Figure 2.15B
Positive Thomas Test

Short hip flexors have been recognized as the most common cause of muscle imbalance, often causing the lower abdominals and hamstrings to lengthen and weaken while the low back muscles shorten and tighten with the hip flexors. This is a frequent finding in golfers with low back pain.

CHAPTER TWO

Supine Knee Extension Test

The Supine Knee Extension Test is used to assess lower hamstring length (Figure 2.16).

Figure 2.16
Supine Knee Extension Test

- Lie on your back with both legs extended on the floor. Place a blood pressure cuff under your lumbar spine at the belly button level. Lift one leg, bent at the knee, until your thigh is at 90° to the floor. Inflate the blood pressure cuff to 70 mmHg. Holding the thigh vertical, slowly extend the knee. Be sure not to lift your head off the floor during the test. Stop as soon as the pressure begins to increase on the gauge.

✓ If you have normal hamstring length, you should be able to straighten your leg to 170°.

!! If you can't straighten your leg to 170°, you have less than normal range of motion in the hamstrings.

▶18 Short hamstrings frequently affect your address posture. When the hamstrings are short, the pelvis cannot rotate forward to maintain an optimal working relationship with the spine. This results in excessive flexion (forward bending) of the low back and often the entire spine. When the lower back and/or middle back are forced to bend more than normal, secondary to short hamstrings, your rotation is limited. Limited rotation reduces your ability to achieve optimal length/tension relationships and lost distance is a common byproduct. Overuse of the arms is common when the spine is limited in its ability to rotate effectively because it is being held in flexion by short hamstrings. This produces similar swing faults as the positive Spinal Rotation Test (Figure 2.14B).

> **WHY USE A BLOOD PRESSURE CUFF?**
>
> The blood pressure cuff is used as a biofeedback mechanism. The inflated cuff under the spine acts as a lumbar support, positioned at the apex of the curvature in this region. A common mistake when performing this test is to let the pelvis roll back and flatten the back as the leg extends, instead of maintaining the lumbar curve and stretching the hamstrings. By using the blood pressure cuff, poor technique is easily identifiable; as the back starts to flatten, the pressure on the cuff will increase, measured by the gauge. The leg is then held in the position at which the pressure on the cuff just begins to rise and the angle measured. The cuff is a very visual aid for correct execution of the test.

Waiter's Bow Test

The Waiter's Bow Test assessess the length of the upper hamstrings and their ability to allow normal motion at the hips.

- To perform the test, stand up straight with good posture and take a pinch of skin at the low back directly opposite your belly button (Figure 2.17). Hold the other arm out like a waiter serving wine and bend forward with the knees locked straight.

- If you have normal upper hamstring length, then you will be able to bend the trunk 50° forward at the hips while holding the pinch of skin, maintaining a curve in the lumbar spine.

- If you do not have adequate hamstring length, then you will not be able to bend forward and maintain your lumbar curve. The pinch of skin will be pulled from your fingers.

- Inability to achieve a normal Waiter's Bow will result in similar swing faults as indicated for the Supine Knee Extension Test (Figure 2.16). The Waiter's Bow Test is also used as the corrective stretch (Figure 3.22).

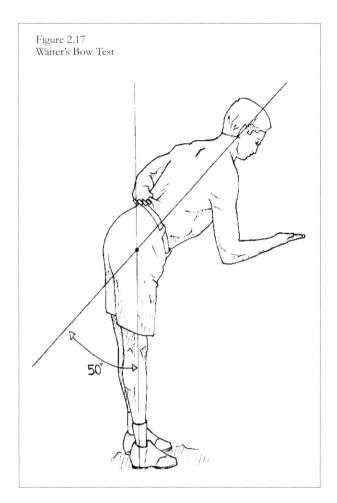

Figure 2.17
Waiter's Bow Test

Cigarette Butt Test

The Cigarette Butt Test is used to assess the flexibility of the internal and external rotators of the hips (Figures 2.18A & B).

☐ To perform the test for the internal hip rotators, stand against a wall with your feet hip-width apart. Imagine you are putting out a cigarette under your right heel by rotating the leg outward, pivoting off the heel. As you turn your leg outward, make sure your pelvis does not move and that your leg is locked straight. To test the length of the external rotators, rotate the leg all the way inward in the same fashion. Make sure your pelvis stays square to the front, not rotating.

✓ Normal internal hip rotators will allow you to rotate the foot outwardly to at least 45° (Figure 2.18A). If you have normal length external hip rotators, you should be able to rotate the leg internally until your foot is inwardly rotated to at least 40° from the starting position (Figure 2.18B).

!! If you cannot inwardly rotate the foot at least 40°, you have limited internal rotation of the hip due to tight external rotators. If you cannot outwardly rotate the foot at least 45°, you have limited external rotation of the hip due to tight internal hip rotators.

▶18 Inability to achieve normal hip internal rotation on the right and/or external rotation on the left will limit your backswing. Reduced external rotation on the right and internal rotation on the left will limit your follow-through. This muscle imbalance often results in overuse of the back and shoulders. Hip rotation imbalance is commonly associated with overuse injury to the back, shoulder and elbows in golfers. This test is particularly important for senior golfers as tight hips encourage lower back pain and power loss.

▶18 Commonly associated swing faults are loss of swing plane, loss of swing axis, loss of swing arc and overuse of arms and wrists. Since the hip joints are major contributors to the swing, motion restriction may be associated with most swing faults.

To correct tight internal rotators (a lack of external rotation), see Figure 3.18. To correct tight external rotators (a lack of internal rotation), see Figure 3.19.

Figure 2.18A
Cigarette Butt Test for Internal Hip Rotators

Figure 2.18B
Cigarette Butt Test for External Hip Rotators

Side Bend Test

The Side Bend Test is used to assess spinal range of motion in side bending (Figure 2.19).

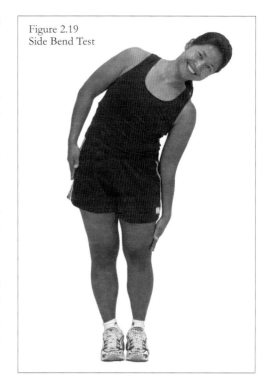

Figure 2.19
Side Bend Test

- ▢ Stand with your feet together and your heels, gluteals, back and head against a wall. Keeping your back, head and shoulders against the wall, slide your hand down the left leg. Go as far as you can without the right heel coming off the ground. Perform the same procedure to the opposite side. Avoid flexion and extension.

- ✓ If you have normal range of motion in your side bend, you should be able to reach the joint line of your knee on both sides equally.

- ‼ Problems to look for are limited range on both sides plus differences in range of motion from side to side. Both sides should be even. If you can't reach your knee on one side, you should stretch the tight side over a Swiss Ball as demonstrated in Figure 3.13.

- ▶18 If you were limited in the Side Bend Test, you were most likely limited in the Spinal Rotation Test (Figure 2.14) as well. This is because spinal rotation and side bending are coupled motions; one cannot happen without the other occuring at the same time. Therefore, restricted side bending will most often be found in conjunction with limited rotation. The swing faults related to limited side bending will be the same as those found with limited spine rotation: excessive sway during the backswing and follow-through, swing faults affecting the swing plane, the clubface angle and maintenance of optimal swing axis, and shoulder compensation with limited coiling.

Arm Raise Test

The Arm Raise Test will assess whether or not you have tight latissimus dorsi muscles.

- The Arm Raise Test is performed by standing with your heels about one foot from a wall. Your buttocks, back and head should all be resting against the wall. Have a partner stick his or her hand between your low back and the wall to determine approximately how much curve you have. Raise your arms in the air and bring them towards the wall, then have your partner re-assess your lumbar curve.

- ✓ If you are able to bring the arms towards the wall without the curve in your lower back increasing and the back moving further away from the wall, you have passed the Arm Raise Test (Figure 2.20A).

- ‼ If you raise your arms in the air and the space between your low back and the wall increases, then the test is positive. This indicates that your latissimus dorsi muscles are shortened (Figure 2.20B).

To correct the length of the muscles, see Figure 3.11.

▶18 A short left latissimus dorsi will disrupt your backswing while shortness in the right latissimus dorsi will distort your follow-through. When a latissimus dorsi is short and becomes **facilitated,** it often over-powers the external rotators of the shoulder. This results in closing of the clubface at impact in most cases. The result of closing your clubface at impact will depend on what type of modifications you have made in your swing plane. Either way, you will have a harder time hitting straight shots with tight latissimus dorsi muscles.

Figure 2.20
Arm Raise Test:
A: Negative / Passed, B: Positive / Failed

GLOSSARY

Facilitated muscles have a lower threshold of stimuli, disrupting synergy of movement and potentially robbing other muscles of neural input.

The Law of Facilitation states that when an impulse has passed once through a certain set of neurons to the exclusion of others, it will tend to take the same course on a future occasion and each time it traverses this path, the resistance will be smaller.

FLEXIBILITY

Thoracic Extension Test

Performed in much the same manner as the Arm Raise Test, the Thoracic Extension Test indicates whether or not there is adequate extension in the thoracic spine to protect your shoulders from injury during a golf swing.

Figure 2.21
Thoracic Extension Test:
A: Good Extension, B: Poor Extension

- To perform the test, start in the same position as the Arm Raise Test with your heels about one foot from the wall and your buttocks, back and head against the wall. If your middle back does not extend adequately and you are unable to rest your head against the wall, you will need to improve your thoracic mobility.

✓ If you can keep your arms in the vertical plane with your butt, back and head on the wall, then you most likely have normal thoracic extension.

!! If the spine does not flatten, then you run the risk of developing a shoulder injury due to lack of thoracic extension.

Thoracic extension is necessary for protecting the shoulder joint from impingement and/or excessive strain once the shoulder reaches 140° of flexion. Inability to fully extend the thoracic spine will significantly hamper your ability to achieve a good backswing or follow-through position. When your backswing position is restricted, you cannot develop a good coil. This will reduce power and distance.

The body usually compensates for reduced thoracic extension by excessive movement of the shoulders and by extending the hips during the backswing. Excessive movement of the shoulders is frequently associated with arm swinging and will eventually lead to shoulder injury. Coming off your swing axis to compensate for a stiff thoracic spine will certainly disrupt both your swing plane and swing arc. Commonly, the result will be a chopping swing with fat, thin and inconsistent shots.

No matter how you slice it, getting the thoracic spine in working order is extremely important to your golf game! The latissimus dorsi, pectoralis minor, rectus abdominus and medial shoulder rotators shorten as the spine increases its thoracic curvature and loses its ability to extend. Foam roller mobilizations demonstrated in Figures 3.28A & B will be a big step in the right direction to reversing this condition. Additionally, single arm rowing exercises (Figure 6.7A and B) will also aid in correction.

McKenzie Press-Up

The McKenzie Press-Up is a mobilization exercise developed by New Zealand physiotherapist Robin McKenzie. It can be used to determine the golfer's ability to extend the lumbar spine (or in layman's terms, to bend the low back backwards) and aid in determining the functional status of the lumbar discs. For the Whole in One Golf Conditioning program, the test has been modified slightly to determine if the golfer has adequate lumbar extension.

- To assess your ability to extend your lumbar spine, lay prone (face down) on the floor with your hands placed just outside the top of your shoulders (Figure 2.22A). Next, inhale deeply and begin pressing your upper body upward as though doing a pushup, but keep your pelvis on the ground (Figure 2.22B). As you push your body upward, exhale. When performing the McKenzie Press-Up, it is very important to relax the glutes and spinal muscles. It is also very important to exhale as you push your body upward, as this facilitates lumbar extension, making the movement easier.

✓ If you have normal spinal extension, you should be able to straighten your arms and keep your pelvis on the floor.

!! The more bend you have in your arms when your pelvis starts to lift off the floor, the greater the restriction you have in your lumbar spine and the more you need to incorporate the McKenzie Press-Up into your stretching program (Figure 3.27).

Figure 2.22
McKenzie Press-Up

FLEXIBILITY

▶18 The golfer lacking lumbar extension will be unable to achieve an optimal backswing or follow-through position. The restriction in lumbar extension may lead to overuse of the shoulders as a compensation, which frequently results in impingement syndrome in the golfer's shoulder. Impingement of the shoulder produces pain when the shoulder is at end range, frequently experienced in the left shoulder during the backswing and the right shoulder during the follow-through.[4]

Note: This exercise is not designed to make the muscles stronger, it is a mobilization exercise to aid in restoring normal motion to the lumbar spine.

WARNING

If you tested positive for either the Thoracic Extension Test or the McKenzie Press-Up, it is essential to have a doctor clear you of ankylosing spondylitis before attempting the mobilizing exercises shown in Figures 3.27, 3.28A & B and 3.29. Ankylosing spondylitis is a fancy name for arthritis of the spine. If it is present, some of the vertebra may be fused together. For those with ankylosing spondylitis, performing these mobilization exercises for the purpose of regaining normal range of motion in extension may result in a fracture of the spine and severe pain. Therefore, make sure your doctor gives you medical clearance to do these exercises!

DEVELOPING YOUR STRETCHING PROGRAM

To develop your stretching program, use Table 2.1. Fill in the ***Your Results*** column with the findings of your flexibility tests. The column on the far right, the ***Corrective Stretch*** column, will indicate which stretch or stretches described in Chapter 3 can be used to restore joint range of motion to normal for that particular test.

Table 2.1

Test	Your Results		Corrective Stretch
	Tight	Normal	
Neck Side Flexion			Neck Side Flexor (Fig 3.2)
Neck Rotation			Neck Rotator Stretch (Fig 3.1)
Sweetheart			Levatator Scapulae Stretch (Fig 3.3)
Apley Scratch			External & Internal Stretch (Fig 3.5 A&B & 3.6)
Pectoralis Major & Minor			Pectoralis Stretch (Figures 3.5B & 3.8)
Spinal Rotation			Trunk Rotation (Figure 3.14)
Thomas Test			Lunge stretch (Fig 3.15) Quad Stretch (Fig 3.25A,B,C)
Supine Knee Extension			Supine Knee Extension Hamstring Stretch (Fig 3.23)
Waiter's Bow			Waiter's Bow Stretch (Fig 3.22)
Cigarette Butt			Internal/External Hip Rotator Stretch (Fig 3.19 & 3.18)
Side Bend			Oblique Abdominal Stretch (Fig 3.13)
Arm Raise			Latissimus Dorsi Stretch (Fig 3.11)
Thoracic Extension			Foam Roller Mobilizations (Fig 3.28 A&B & 3.29)
McKenzie Press Up			McKenzie Press Up (Fig 3.27)

IF YOU NEED HELP!

By using the Whole In One Golf Conditioning assessment techniques demonstrated in this chapter, you will be able to do a thorough job of correcting your muscle imbalances. C.H.E.K Institute-trained professionals can perform a comprehensive assessment that takes approximately three hours. As you can imagine, this type of assessment exposes any and every possible limiting factor that may prevent you from playing or learning to play your best golf. If for any reason you do not feel comfortable performing your own assessment or if you feel you have not achieved the results you were after by working on your own, then call the C.H.E.K Institute and make an appointment for a full Whole In One Golf Conditioning assessment and exercise program. Alternatively, to find a C.H.E.K Institute-trained professional call 1.800.552.8789, or search the online database at **www.CHEKconnect.com**.

CHAPTER 3

STRETCHING

Prior to beginning any stretching routine, it is important to be warm. When stretching, you are lengthening muscles that may be very tight. If your body is cold, your chances of over-stretching and injuring a muscle are significantly increased. Therefore, on a cold day it is important to find a warm place to stretch. When tension is placed on a musculotendinous unit, heat is generated. The heat generated by the tension on the muscle-tendon unit breaks down the ground substance in the connective tissue housing the muscle fibers, allowing the musculotendinous unit to elongate. If you are cold, this process cannot happen as effectively.

For optimal results with your stretching, it is important to wear loose clothing that will insulate your body, retaining the heat generated while stretching. If your body gets cold, you can lose your stretch before you get to the first hole!

You may be feeling overwhelmed by all the stretches and information to this point. Don't worry, there is a system to using all this information. It is unrealistic to expect you to do all the stretches described in the book even though many of you would benefit from doing exactly that. To assist you in selecting the stretches most likely to help you in your quest for optimal golf performance, a table of common golf faults has been provided. Some of these faults can be significantly improved upon through corrective stretching (Table 3.1, page 67).

The best way to use this table is to find your most prominent technical fault on the list. Any stretch listed in the table for your prominent swing fault that was not previously identified by length tension tests should be added to your stretching program. Performing this combination of stretches will serve as an integral part of correcting pending biomechanical and swing faults. The stretches should be performed as developmental stretches until the fault is corrected. After correction, you should perform a maintenance program to prevent the fault from returning.

> *If your body is cold, your chances of over-stretching and injuring a muscle are significantly increased. When you are preparing to play golf on a cold day, it is important to find a warm place to stretch.*

CHAPTER THREE

If there are stretches identified for your fault that also appear on the list of tonic or hyperactive muscles shown in Table 2.1, they should be performed prior to warming up to practice, or playing a game of golf, as well as part of your developmental program.

Neck Rotators

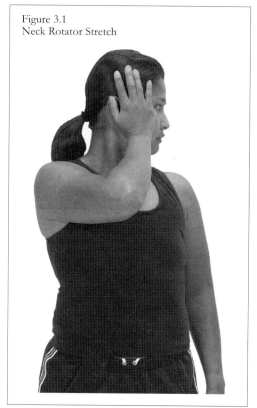

Figure 3.1
Neck Rotator Stretch

- Maintain an upright posture, either sitting or standing.

- Rotate your head to one side.

- Place the opposite hand on your cheek and face, creating an immovable barrier. Be careful not to force your jaw laterally during the contraction phase.

- Inhale and gently rotate your head into your hand while looking in the direction of the rotation towards the head for five seconds.

- After five seconds of gentle pressure (1 lb.), exhale as you look over your shoulder and rotate your head into the stretch.

- It is important to rotate into the stretch immediately upon taking the pressure of rotation from your hand.

- This procedure should be repeated three to five times in each direction or until you no longer feel your range of motion increasing.

!! If you find that you rotate noticeably further in one direction than the other after stretching both sides, it is important to perform the Sweetheart Test (Figure 2.10). A positive Sweetheart Test indicates levator scapulae tightness, which can cause asymmetrical rotation of the neck. If you have a positive Sweetheart Test, you should have an orthopedic physical therapist, chiropractor or a C.H.E.K Practitioner assess your neck mechanics.

Neck Side Flexors

- To perform this stretch you need to anchor the shoulder girdle and provide a foundation from which to stretch. Sit or stand with good postural alignment. If seated, grasp the right side of the chair or bench with your right hand. If standing, depress your shoulder by tightening your latissimus dorsi muscles. To do this, reach down the side of your leg as though delving into an endless pocket. Do not lean the body to that side, just depress the shoulder as far as you comfortably can without actually leaning.

- Use the opposite hand to gently grasp the side of your head.

- Carefully draw the neck away from the anchored shoulder until the muscles feel a comfortable stretch.

- Inhale deeply and side bend the head up into your hand but do not let your head move. Hold the head still as you put about 1 lb. of pressure into your hand for five seconds.

- After five seconds, exhale and immediately depress the shoulder and, using the arm, gently draw the head and neck into side flexion.

- Repeat this process three to five times on each side.

Figure 3.2
Neck Side Flexor Stretch

Levator Scapulae

- Stand sideways about 1ft. from a wall.

- Reach up with the arm closest to the wall.

- Looking away from the wall, bend this arm and grab the base of your skull.

- Lean toward the wall, allowing the hand holding your skull to transfer the lean into the neck, creating a side-bending action on the neck. BE VERY CAREFUL NOT TO BE TOO FORCEFUL!

- Once your neck has a gentle stretch, simultaneously inhale and press your elbow into the wall and your head into your hand for five seconds.

- After five seconds, exhale and relax toward the wall, allowing your arm to gently side-bend your head and neck away from the wall to the point at which you feel a stretch.

- Repeat this process three to five times on each side.

  !! If you passed the Sweetheart Test (Figure 2.10) on one side and not the other, perform this stretch only on the side that tested positive for levator scapulae tightness. If you stretch both sides, you will not improve the imbalance.

  It is important to keep the levator scapulae flexible because it has the ability to significantly alter both swing mechanics and neck and shoulder function!

Figure 3.3
Levator Scapulae Stretch

Neck Extensors

- Maintain an upright posture, either sitting or standing and let your head drop toward your chest.

- With one hand placed on the back of the head, take a deep breath and extend your neck into your hand with 1-3 lbs. of pressure, without letting your head actually move.

- After five seconds of pressure, relax as you exhale and use the hand on the back of your head to gently draw your head forward into flexion.

- Repeat this procedure three to five times.

- !! If you have any pain during this stretch, particularly pain radiating toward the shoulder or into the arm, stop the stretch immediately. Pain into the shoulder or arm while performing this stretch may indicate cervical disc pathology and should be cleared by your medical doctor, an orthopedic physical therapist or chiropractor.

Figure 3.4
Neck Extensor Stretch

Medial Shoulder Rotators & Pectoralis Minor

The medial rotators of the shoulder may be stretched in a doorway or on a Swiss ball (Figures 3.5A & B). The pectoralis minor muscle works in conjunction with the medial shoulder rotators, therefore by stretching one, you will also be stretching the other

- Stand in a doorway and place your arm in the same position as you would if you were about to throw a ball.

- Place your forearm and hand in front of the doorjamb.

- Gently rotate your trunk forward around your arm as though your arm were a stationary object.

- Once you have reached a position in which you feel a stretch on the medial rotators of your shoulder (the front area), take a deep breath and press your hand into the doorjamb.

- Hold about 1 lb. of pressure on the doorjamb for five seconds.

- Exhale and rotate the trunk around the arm, increasing the stretch.

- This should be repeated three to five times.

Figure 3.5A
Medial
Shoulder
Rotator Stretch
Against the
Wall

Using the Swiss ball is another excellent way to stretch your medial shoulder rotators and pectoralis minor.

- Place your hand and arm over the apex of a Swiss ball with your shoulder resting on the ball as seen in Figure 3.5B.

- Allow your body to drop forward while allowing your shoulder blade to move toward your spine.

- Once you feel a stretch, inhale and press your hand and shoulder into the ball for five seconds.

- Relax and allow your body to drop forward, bringing your shoulder blade closer to your spine.

- As you can see by looking at Figure 3.5B, the apex of the ball will be progressively higher than the plane of the shoulders.

- The stretch should be repeated three to five times on each side.

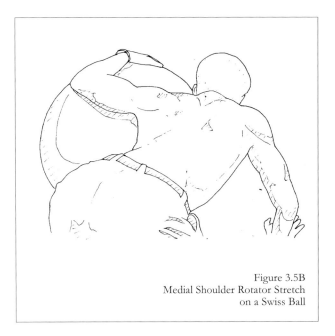

Figure 3.5B
Medial Shoulder Rotator Stretch
on a Swiss Ball

!! **Note**: It is very important that you not perform these stretches if they cause any discomfort in the shoulder. You may find that the doorway stretch is uncomfortable, but not the Swiss ball one. If this is the case, continue only with the Swiss ball stretch. If both of the stretches bother your shoulder(s), consult an orthopedic physician. Many athletes are loose in the front region of their shoulder joints from past injury and from lifting weights with poor technique. This stretch may serve to stretch areas that are already too loose.

CHAPTER THREE

Lateral Shoulder Rotators

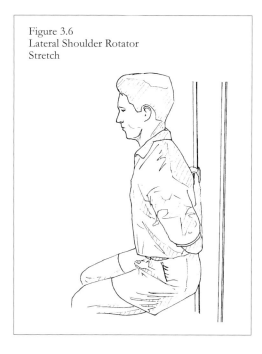

Figure 3.6
Lateral Shoulder Rotator Stretch

- Reach behind your back and grab a doorknob (Figure 3.6) or any object sufficiently solid to serve as an anchor point.

- Slowly lower your body to the point at which you feel a stretch behind your shoulder.

- Inhale and gently pull downward on the doorknob for five seconds.

- Exhale and allow your body to slide down just enough to comfortably increase the stretch.

- Repeat this process three to five times on each side.

Rhomboids

The rhomboids and related muscles are used to bring the shoulder blades toward the midline. They can be stretched quite well with the use of a Swiss ball.

- Kneel in front of a Swiss ball and place your elbow on the ball.

- Draw the arm across the body as it rests on the ball by rotating the ball in the direction of the arrow.

- When the muscles have been comfortably stretched, inhale and activate the muscles as though to pull the shoulder blade toward the spine. Putting some pressure on the ball with the elbow will assist in activating the correct muscles.

- Hold this pressure for five seconds.

- Release the pressure as you exhale and move into the next stretch position.

- Repeat the cycle three to five times on each side.

Figure 3.7
Rhomboid Stretch

STRETCHING

Pectoralis Major

The pectoralis major (chest muscle) is stretched in much the same way as the medial shoulder rotators, with the exception being that the emphasis is on taking the arm away from the midline, not so much on moving into lateral rotation. Again, a doorway or a Swiss ball may be used.

In a Doorway

- The doorway stretch is performed in the same position as in Figure 3.5A.

- The emphasis is moving into the doorway while using the hand and forearm as the stationary point from which you stretch.

- Once you feel a stretch on the chest, inhale and press the forearm into the doorjamb for five seconds.

- As you relax, exhale and allow the body to move further into the doorway, placing additional stretch on the muscle.

Using a Swiss Ball

- Stretching the chest on the Swiss ball (Figure 3.8) is again performed in much the same manner as the medial shoulder rotators were stretched on the Swiss ball.

- With the forearm on the ball, keep the shoulders parallel to the ground as you allow the arm to be stretched back by dropping the body forward.

- Upon reaching the point of a comfortable stretch, inhale and press the forearm into the ball for five seconds.

- After five seconds, exhale as you relax and move immediately into a new stretch position.

- Perform this sequence three to five times on each side.

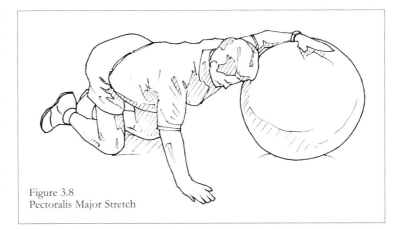

Figure 3.8
Pectoralis Major Stretch

PECTORALIS MINOR

See Medial Shoulder Rotators and Pectoralis Minor stretch, Figures 3.5A and 3.5B.

Wrist Extensors

- Reach either over or under the arm to be stretched and grasp the fingers and hand to be stretched.

- Holding the arm to be stretched straight, flex the wrist and put a slight medial rotation on the arm to increase the stretch to the wrist extensors.

- Once a comfortable stretch is achieved, inhale and attempt to extend the wrist against the stabilizing hand for five seconds.

- Immediately upon relaxing, exhale and use the working hand to increase the medial rotation of the arm and flexion of the wrist, increasing the stretch on the wrist extensors.

- This stretch should be repeated three to five times on each arm.

Wrist Flexors

- With the arm held straight out in front of you, draw the wrist and fingers back to the point at which you feel a comfortable stretch (Figure 3.10).

- Inhale and activate the wrist and finger flexors of the arm being stretched by pushing the fingers into the opposite hand for five seconds.

- After five seconds of moderate pressure, exhale as you relax and simultaneously increase the stretch.

- Repeat the process three to five times on each side.

Figure 3.9
Wrist Extensors Stretch

Figure 3.10
Wrist Flexors Stretch

Latissimus Dorsi

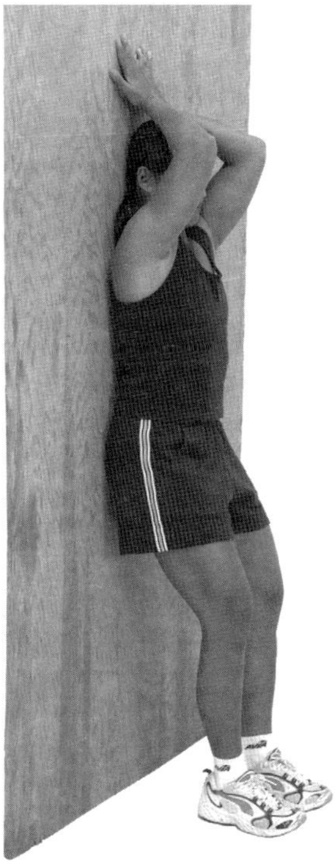

- Lean against a wall or doorjamb with your feet about 1 ft. from the wall and your knees slightly bent.

- Bring your hands up in front of your face as though you were looking at your palms.

- Pull your elbows inward until they are in line with or slightly inside your shoulder joints.

- Draw your navel inward toward your spine as you roll the pelvis backward, flattening your back against the wall. It is important to keep your head and as much of your spine on the wall as possible.

- Raise your hands over your head, trying to touch the palms of your hands against the wall while keeping your elbows within the shoulder joints.

- Hold the stretch for twenty seconds.

- For best results, this stretch should be performed three times per session.

Figure 3.11
Latissimus Dorsi Stretch

Alternative Method

Another method of performing the stretch is to do it in the reverse order. This stretches the latissimus dorsi from the bottom up, instead of from the top down as shown above.

- Assume the same start position against the wall, but this time keep a natural curvature in your lower back.

- Keep the elbows inside the shoulder as you raise your hands up to the wall.

- Once your hands are touching the wall, draw your elbows close together and attempt to roll the pelvis backward, flattening your spine against the wall, or as much as you can comfortably tolerate for twenty seconds.

CHAPTER THREE

Rectus Abdominis

The most effective way to stretch your rectus abdominis, the muscle that has the washboard appearance in body builders and lean athletes, is to lay over a Swiss ball (Figure 3.12). Laying over a Swiss ball not only stretches your abdominals, it also helps restore your spine's ability to extend (bend backwards). This is important because a lack of extension in the middle back is a common contributor to shoulder pain and swing faults in golfers.

- Sit on a Swiss ball, then walk your legs out and roll backwards until you are lying over the ball, with your arms extended over your head.

- You can increase the stretch by slowly straightening your legs, which will push your upper body further over the ball.

- Alternate moving slowly forward and backward over the ball, bringing your hands closer to the floor and then taking them away.

- The stretch is best performed for a minimum of one minute.

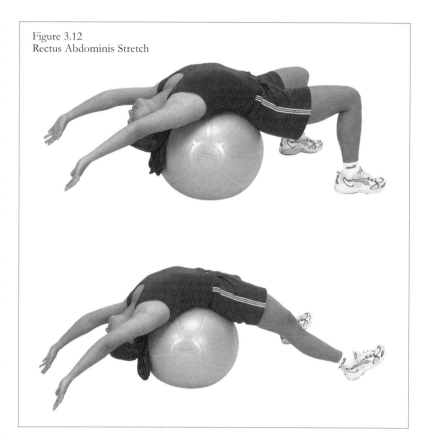

Figure 3.12
Rectus Abdominis Stretch

A lack of extension in the middle back is a common contributor to shoulder pain and swing faults in golfers.

SPECIAL NOTE

If you experience dizziness when looking at an airplane flying by or when putting things away in high cupboards, you may experience dizziness when stretching your abdominal muscles on a Swiss Ball. It is very important that you stop the stretch immediately if you feel any unusual symptoms. Those not having sufficient spinal extension may compensate by over-extending their necks, which can result in symptoms such as nausea, dizziness or changes in vision. Any unusual symptoms indicate the need to see your doctor for a complete evaluation of your neck to rule out occulsion of the vertebral artery. Should your neck be cleared as normal, you should have your inner ear assessed by a physician specializing in the ear, nose and throat.

Oblique Abdominals

The oblique abdominals, or the muscles on the side of your trunk, can also be stretched very well on the Swiss ball.

- To get properly positioned, roll down the ball from a sitting position into a supine position and then carefully roll onto your side. You may feel more comfortable using a spotter for this stretch.

- Use a wall or partner to anchor your feet.

- Grasp the wrist of the top hand with the bottom hand as seen in Figure 3.13.

Figure 3.13
Oblique Abdominal Stretch

- Gently begin rolling the upper body forward while putting a mild downward tug on the upper arm.

- When you feel an area of tightness, hold that position until the tightness eases.

- Progressively roll forward and backward looking for tight areas, being careful not to fall off the ball. Again, get a spotter if you are at all likely to have a balance problem. You may also hold onto a stable object until you feel more comfortable with the stretch.

Trunk Rotation

The trunk rotation stretch is extremely valuable to the golfer with reduced rotational flexibility. Golf is very dependant upon spinal rotation and any limitation will certainly hinder your golf swing.

- Lie on your back as seen in Figure 3.14. The hips should be flexed until the knees point to the ceiling, making sure the lower legs are completely relaxed.

- Place the arm of the side you are going to rotate your lower body toward on your thigh, so it can aid in lowering your legs.

- Stretch the other arm out on the floor for stabilization.

Figure 3.14
Trunk Rotation Stretch

CHAPTER THREE

- Slowly let the legs roll to one side until you feel a comfortable stretch in your back or side.

- Hold for five seconds, then roll to the other side.

- Continue to alternate sides until you no longer make progress or until your thighs are on the ground with your hips flexed to 90°.

- Generally, ten repetitions to each side should be performed.

- To stretch higher up the spine, draw the knees toward the chest slightly. As the hips flex, the stretch moves progressively upward as you rotate.

  - ▶18 Normal spinal rotation is achieved when your legs lie flat on the floor and your opposite side shoulder remains on the floor. As mentioned in Chapter Two, achieving normal spinal rotation is of great benefit to any golfer.

Hip Flexors

The lunge stretch is used to lengthen the hip flexors.

Figure 3.15
Lunge Stretch

- Assume the lunge position as seen in Figure 3.15. and draw the navel inward.

- A) Roll the pelvis backward (which will flatten your low back) and B) begin to move the whole pelvis forward, keeping it square to the front. It is not uncommon for the pelvis to swing outward toward the trailing leg side, which is an indication of very tight hip flexors.

- C) The stretch can be increased by reaching the arm on the trailing leg side over your head and side bending the trunk away.

- An additional stretch can again be added by slightly side bending the torso toward the front leg side while maintaining the previous pelvis and arm position.

- The lunge stretch should be held for twenty seconds on each side, alternating between left and right sides three times. The only exception is when you have only one side testing positive on the Thomas Test (Figure 2.15B.). In this case, you need only stretch the tight side until balance is achieved.

Lumbar Erectors

- To stretch the lumbar erectors, also known as the low back muscles, lie on your back and pull both legs toward your chest as seen in Figure 3.16. To get the best stretch, pull the legs toward the chest as you exhale.

- Once you feel a comfortable stretch in your low back, hold your legs with your arms, resisting as you gently push your legs into your hands for five seconds. Exhale and immediately draw the legs a little closer to the chest.

- Take a breath and exhale, then activate the low back and hip muscles for another five seconds. Repeat the process three to five times per leg.

If you find the double knee to chest stretch uncomfortable, you may use the single knee to chest stretch instead (Figure 3.16B). This may be the case for someone with a herniated disc or for those of you who experience pain with flexion.

Figure 3.16A
Lumbar Erector Stretch: Double Knee to Chest

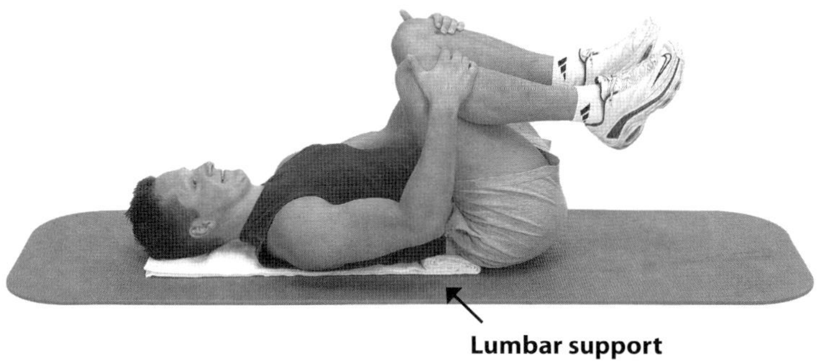

Lumbar support

Figure 3.16B
Lumbar Erector Stretch: Single Knee to Chest

SPECIAL NOTE

If you have been diagnosed with a lumbar disc bulge, have your doctor approve the use of this stretch. If you have any discomfort with the stretch as suggested above, then it is recommended that you perform the stretch with a lumbar support as seen above in Figure 3.16. To make a lumbar support, roll a bath towel to the width of your hand and when compressed, the thickness of the fattest part of your hand. The towel support should be placed at belt line level or in the space just behind the belly button (L3), which is at the same level as the apex of your lumbar curvature. Using the towel support will significantly reduce the stretch on the lumbar erectors while improving the stretch on the gluteus maximus and upper hamstrings. This modification often reduces tension on the low back reflexively. It is very important that you not let your sacrum rise off the floor during the stretch.

90/90 Hip Stretch

The 90/90 Hip Stretch may be the single most effective stretch a golfer can do if he or she has less than ideal hip rotation during the swing.

Figure 3.17A
90/90 Hip Stretch:
Side View

- Sit on the floor with both your front and back legs bent to 90°. The angle in the groin created by both legs should also be 90°, as seen in Figure 3.17B.

- Take the hand on the same side as the forward leg and place it on the ground next to your hip, with the inside of the arm facing forward. This hand placement will help hold the correct position as you perform the stretch.

- Tip your pelvis forward as though it were a bowl and you were trying to pour the contents out over your belt buckle. Hold this position. Done correctly, you will have increased the curvature of your lumbar spine.

- Inhale as you bend forward, keeping the curve in your low back, bending from the hip.

- To aid in holding the proper stretch position, put pressure into the hand by the hip.

- As you lower your trunk toward the floor, allowing yourself to bend only from the hip, keep the chest up, shoulders parallel to the floor, and the eyes level with the horizon (Figure 3.17A).

- At the point when you feel a comfortable stretch, hold the position and press the front knee and ankle firmly into the ground for five seconds.

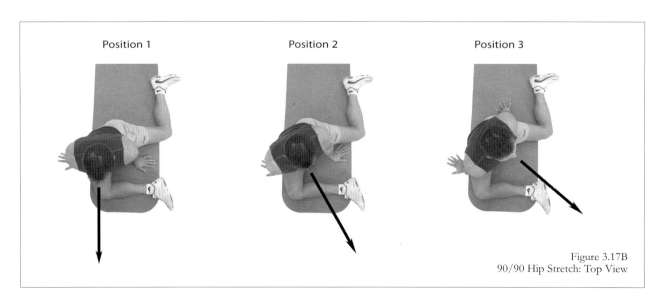

Figure 3.17B
90/90 Hip Stretch: Top View

- Relax, exhale and move further into the stretch. Repeat the process three to five times on each leg.

- Next, move your torso so that your head lines up with the middle shin and repeat the process (Figure 3.17B, Position 2). As you become more flexible, you can attempt to perform the stretch with the head aligned with the front foot (Figure 3.17B, Position 3). Perform this process on each side until you feel you have reached your stretch potential for that session.

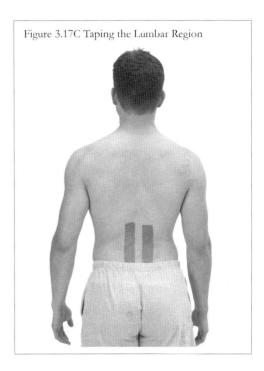

Figure 3.17C Taping the Lumbar Region

Note: If you are tight in the hips you will find it hard to keep the lumbar curve and pelvis tilted forward as described above. What you can do to help maintain the curve in the lumbar spine is to tape your low back as a stretching aid. This is done by having someone run a strip of athletic tape from the level of the bottom rib to the sacrum while you are standing with good upright posture (see Figure 3.17C). After being taped, perform the stretch exactly as outlined here. If you feel the tape pull as you lean forward into the stretch, this indicates that you are not holding the proper spine position and that you need to tip your pelvis forward to reduce the tension on the tape.

This modification is very important for anyone with a history of back pain because it prevents you from over-stretching the low back during what is intended to be a hip stretch.

Cigarette Butt Stretch for External Hip Rotators

The internal and external hip rotators are commonly overlooked in many stretching programs. The golfer can't afford this oversight, as flexible hip rotators are key in the execution of a good golf swing.

- Stand with your feet parallel and slightly farther apart than the length of your foot. Turn the foot of the leg you wish to stretch inward, pivoting off the heel as though you were putting out a cigarette. Make sure to keep your pelvis square to the front.

- Once you have turned the foot inward as far as possible, hold the foot pressed into the ground with the knee just slightly unlocked but held stiff.

- Turn your pelvis in the opposite direction of the foot movement without allowing any movement of the inwardly rotated foot (Figure 3.18).

- Once you feel a comfortable stretch in the hip, hold the pelvis in that position, inhale and gently attempt to externally rotate the leg against the ground, as though you were trying to take it back to the starting position. Do not actually let the foot or leg move, just activate the muscles. Hold for five seconds.

- Exhale as you relax and move the pelvis toward the hip being stretched to a new stretch position. This process should be repeated three to five times on each leg.

Figure 3.18
Standing External Hip Rotator Stretch

Standing Internal Hip Rotators

- To stretch the internal rotators of the leg, the process is much the same as the external rotators, except you now rotate the foot outward, pivoting off the heel.

- Rotate the leg outward as far as possible, keeping the hips square to the front.

- Then hold the leg still and rotate the pelvis inward, opposite the movement of the foot.

- Once you have a good stretch in the internal hip rotators, take a deep breath. Keep the knee rigid and the leg motionless as you attempt to internally rotate the leg against the ground for five seconds.

- Exhale and move the pelvis until a new stretch position is reached.

- The process should be repeated three to five times per side.

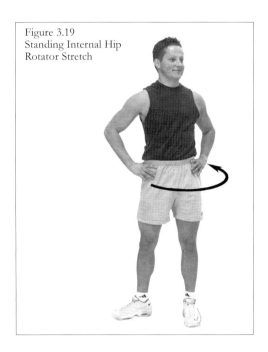

Figure 3.19
Standing Internal Hip Rotator Stretch

Groin Stretch (Indian Sit)

The Indian Sit Stretch is a classic groin stretch, used by many athletes.

- Sit on the floor in an upright posture with the feet together and the heels drawn into the groin as far as possible. The forearms should rest along the shins with the hands grasping the ankles as seen in Figure 3.20.

- Inhale while using the arms to open the legs to the point of a comfortable stretch.

- Activate the groin muscles for five seconds, attempting to close the legs against the arms, which provide an immovable barrier.

- Exhale and use the arms to take the legs to a new stretch position. It is very important not to reduce the stretch as you stop the contraction. When you relax, you must immediately move to the next stretch position or you will lose the nervous system inhibition created by the contraction.

- The stretch should be performed for three to five repetitions.

Figure 3.20
Indian Sit (Groin) Stretch

Groin Rocking Stretch

The Rocking Groin Stretch is a very effective groin stretch and should be performed on a mat, thick carpet, or grass to protect the knees.

- Assume a kneeling position and spread the knees as far as you comfortably can (Figure 3.21A).

- Rock forward (Figure 3.21B) and rotate your pelvis in the direction of the arrow until you feel a comfortable stretch in the upper groin area.

- Breathe in and gently squeeze your knees into the ground for five seconds.

- Exhale and relax the groin as you sink forward into a new position.

- The stretch should be performed from the back, middle and forward positions until the groin is loosened.

Unlike other stretches in this book, the Rocking Groin Stretch should be done at many angles and as many times as needed to loosen the groin. The stretch can also be enhanced by slightly rotating the pelvis to one side or the other once in the stretch position.

> P18 The groin muscles have a large and powerful connection to both the pelvis and the femur (the upper leg bone). It is important to stretch the groin well when it is tight, or holding a stable swing axis will be very difficult. A tight groin can easily disrupt pelvic rotation during the swing.

Figure 3.21A
Groin Rocking Stretch: Start

Figure 3.21B
Groin Rocking Stretch: Finish

Standing Hamstring Stretch (Waiter's Bow)

- Stand with your feet parallel and close together (Figure 3.22). Keep your legs straight and stick your buttocks out until you have an arch in your low back.

- As you bend forward from the hips, do not let your low back begin to round. Try to imagine that you are a waiter at a classy restaurant serving expensive wine.

- Bend forward from your hips until you feel a comfortable stretch on your hamstrings. The stretch may be felt behind the knee or below the buttocks depending on where you are the tightest.

- Hold the stretch position for twenty seconds.

- Relax and come to an upright standing position for a second or two and then repeat the stretch. This process should be repeated at least three times or until your hamstrings no longer feel as though they are loosening.

Figure 3.22
Waiter's Bow Stretch

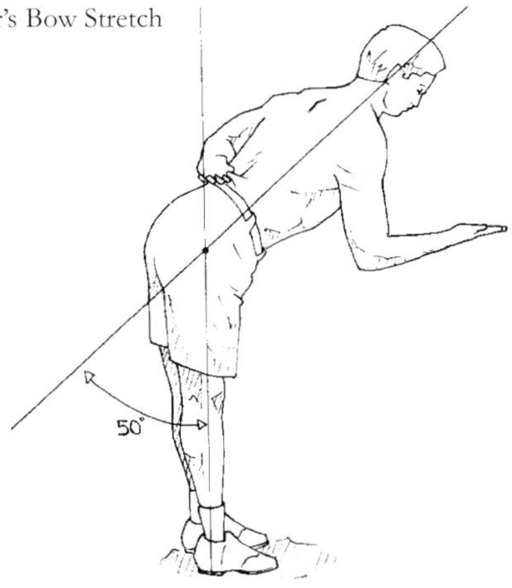

> **Note:** If you have a hard time holding the arch in your back, apply tape as demonstrated in Figure 3.17C. When you feel the tape pull, make an effort to increase your lumbar curvature.

Supine Knee Extension Hamstring Stretch

The Supine Knee Extension Stretch is particularly effective for stretching the hamstring muscle at the knee. This stretch is useful for people who sit at work, as they often have hamstring tightness in this pattern of movement.

- Perform the stretch from a supine (on your back) position with a towel rolled up and placed under the lumbar spine at the belt line level (Figure 3.23). The towel (when compressed) should be the width and thickness of the fattest part of your hand.

- Grab one thigh just below the knee and bring the bent leg up until the thigh is perpendicular to the floor.

- Pull your toes back toward your shin as far as you can and then slowly straighten your leg without letting the thigh move in your hands.

- When you reach the point of a comfortable stretch, hold it for twenty seconds and then switch sides. Alternate between left and right sides three times.

Figure 3.23
Supine Knee Extension Hamstring Stretch

Supine Quadriceps Stretch

- From a kneeling position, drop back onto your hands as seen in Figure 3.24. You may need a mat to perform this stretch to avoid discomfort in the knees.

- Draw the navel inward and roll the pelvis under.

- Project the hips forward as though you were trying to touch the ceiling with your belt buckle.

- Once you have a comfortable stretch, hold it for twenty seconds.

- This stretch should be repeated three to five times or until you no longer feel you are making progress.

Figure 3.24
Supine Quadriceps Stretch

The Supine Quadriceps Stretch is particularly effective for stretching the upper portions of the quadriceps where they attach to the pelvis, as well as stretching the hip flexor muscles. To get a better stretch of the quadriceps at the knees, try the Swiss Ball Quadriceps Stretch in Figure 3.25A, B & C.

Swiss Ball Quadriceps Stretch

The Swiss ball is an excellent stretching aid, although initially the Swiss Ball Quadriceps Stretch may be a little challenging.

- Begin in a sprinter's start position as seen in Figure 3.25A, with the foot and ankle of the leg to be stretched on the ball.

- Slowly rise upward (Figure 3.25B).

- For added stability and control over the quadricep being stretched, place the hand of the side being stretched on the ball adjacent to your foot (Figure 3.25C). If you have a hard time getting into the stretch position, try using a smaller Swiss Ball.

- From this upright position with one foot and one hand on the ball or hip, draw your navel inward as you roll the pelvis under, flattening the back. This will increase the stretch on the quadriceps.

- If you can comfortably handle a greater stretch, continue by projecting your pelvis forward from the tail-tucked position.

STRETCHING

Figure 3.25B
Swiss Ball Quadriceps Stretch: Finish

Figure 3.25C
Swiss Ball Quadriceps Stretch: Supported

- Stretch each leg for twenty seconds, alternating left and right three times.

- The Swiss Ball Quadriceps Stretch can also be done in the contract/relax format. This technique, which is often more efficient, requires that you move comfortably into the stretch and then press your foot into the ball for five seconds. Relax for five seconds and increase the stretch. Repeat this process three to five times on one leg before moving to the opposite leg.

Calves

The calves consist of two muscles that frequently need stretching, the bulky gastrocnemius and the long slender soleus muscle.

- To stretch the gastrocnemius, stand on the edge of a step, curb or similar stable object and allow the heel to drop toward the floor while maintaining a straight leg (Figure 3.26A).

- Hold the stretch for twenty seconds and then stretch the other side. Alternate back and forth three times for best results.

- To stretch the long soleus muscle, simply bend your knee slowly from the Gastrocnemius Stretch position. You will feel the stretch migrate down your calf muscle toward your Achilles tendon (Figure 3.26B). Do not allow the stretch to move into the tendon.

- Hold this position for twenty seconds. Alternate between the left and right leg three times.

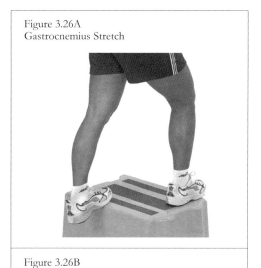

Figure 3.26A
Gastrocnemius Stretch

Figure 3.26B
Soleus Stretch

McKenzie Press-Up

- To perform this excellent extension mobilization exercise for the low back, lie prone (on your stomach) on the floor. Place your hands just outside the tops of your shoulders.

- Inhale deeply and begin pressing your upper body upward as though doing a push-up, but leave your pelvis on the ground. As you push your body upward, exhale to facilitate lumbar extension. It is very important to relax the buttocks and spinal muscles.

- Hold the position at the top until you need to take a breath.

- As you inhale, slowly lower your body to the floor. This process should be repeated ten times per stretching session.

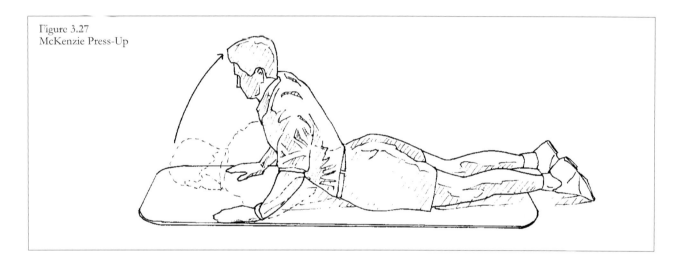

Figure 3.27
McKenzie Press-Up

> **SPECIAL NOTE**
>
> This is not a strengthening exercise. It is a mobilization exercise to aid in restoring normal motion to the lumbar spine. It is not uncommon to feel some discomfort at the end range of movement, particularly if you are limited in range of motion. It is also normal to feel some mild discomfort for the first 3-5 repetitions, which should progressively decrease between about the fifth to tenth repetition. If you have discomfort with every repetition, you may consider consulting a certified orthopedic physical therapist or your physician.

Thoracic Spine – Foam Roller Mobilizations

> **Note:** Prior to mobilizing your spine with a foam roller, it is very important that you get approval from your doctor. As mentioned in Chapter Two, some people have arthritis of the spine that precludes them from performing this mobilization.

- Place a 3" or 4" foam roller* perpendicular to your spine at the point slightly below your shoulder blades or where you feel the restriction begins. The point of restriction can be identified by raising your arms in the air and having someone tell you where there is still a hump in your back.

- Cradle your neck with your hands (Figure 3.28A). Make sure that you do not hold your head, or the mobilizations may cause neck discomfort.

- With your neck supported, begin mobilizing your spine by inhaling as you allow yourself to drop backward over the roller toward the floor (Figure 3.28B). Limit your inhalation to the point just prior to your spinal extensors, tightening, which is typically the last 1/3 of the inhalation. This will allow the mobilization to be more effective.

- Hold the end point for three to five seconds then exhale and come back to the start position. Repeat three to five times.

Figure 3.28A
Thoracic Spine Mobilization: Start

Figure 3.28B
Thoracic Spine Mobilization: End

- Move the foam roller up or down one vertebral segment and repeat.

- The mobilizations should be done before golf, exercise or in the middle of the day and in the evening. Two times per day is adequate to improve spinal flexibility. Should you feel discomfort in the spine after performing the mobilizations, it is likely that you are doing too much. Reduce the number of repetitions. If you still feel pain, consult your doctor.

*See Resources section on page 219.

It is often helpful to lay on the roller lengthwise in the middle or at the end of each day (Figure 3.29). This allows gravity to stretch the ligament in front of your spine, which helps restore optimal spinal curvatures. The better your alignment, the more efficiently you rotate during the golf swing.

- Place the roller along the spine from the base of the skull to the tail bone. The knees should always be bent and the feet flat on the floor.

- Start with five minutes a day and work up to fifteen minutes a day.

- To improve the rotational mobility of your spine, allow your pelvis and shoulders to roll in opposite directions on the roller as shown in Figure 3.29. Repeat this process until you feel loose.

Figure 3.29
Longitudinal Mobilizations

FAULTS AND FIXES

FAULTS	CORRECTIVE ST[...]
1. Poor posture at address	3.2-3.5, 3.15, 3.22, 3.26, 3.2[...]
2. Loss grip during swing	3.1 – 3.10
3. Misalignment of clubface at address	3.1 – 3.10, 3.13 – 3.15
4. Misalignment of body at address	3.1 – 3.29 Test
5. Stance too wide	3.17, 3.18, 3.22
6. Tension at address	3.1 – 3.29
7. Clubhead moves inside too quickly	3.1, 3.3, 3.7, 3.8, 3.10, 3.11, 3.13, 3.14
8. Taking the clubface back closed	3.1, 3.5R, 3.6L, 3.7L
9. Flat laid-off backswing	3.11, 3.12
10. Body turn completed too early	3.17 – 3.21, 3.24, 3.25
11. Incomplete body turn	3.11 – 3.16
12. Overswinging – bend in left arm	3.3, 3.5R, 3.6L, 3.7L, 3.17 – 3.21, 3.24, 3.25
13. Wrong plane – too upright or too flat	See Faults 7, 8, 9 and 11
14. Club jammed behind body	3.6R, 3.7R, 3.11R, 3.13, 3.14, 3.17L, 3.18L
15. Poor weight transfer – no legs	3.15, 3.17, 3.18, 3.19, 3.20, 3.21, 3.22, 3.24, 3.25
16. Early release – casting the clubhead	3.17 – 3.21, 3.24, 3.25
17. Dominant left side – poor body release	3.11 – 3.21, 3.24, 3.25
18. No extension past impact	3.11 – 3.15, 3.17 – 3.19, 3.20R, 3.25R
19. Poor follow-through position	3.11 – 3.15, 3.17 – 3.19, 3.20R, 3.25R
20. Excessive head movement	3.1 – 3.6, 3.17 – 3.21, 3.24, 3.25
21. Slices and pulls	3.3, 3.5, 3.6, 3.17 – 3.21, 3.24, 3.25
22. Hooking and pushing	3.3, 3.5R, 3.7R, 3.10R, 3.11R, 3.17 – 3.21, 3.24, 3.25
23. The Shank	3.1 – 3.29 Test
24. Topped-skulled – fat-heavy shots	3.1 – 3.29 Test
25. Putting – excessive body motion/poor use of spinal axis	3.17 – 3.21, 3.24, 3.25
26. Females – poor power development	3.3, 3.14, 3.15, 3.16
27. Seniors – loss of mobility	3.1 – 3.29 Test
28. Seniors – loss of power	3.1 – 3.29 Test

DEVELOPMENTAL STRETCHING

Developmental stretching, also referred to as "corrective stretching," is needed by those who have muscle imbalances and joint restrictions that could potentially cause altered swing mechanics. This type of stretching corrects shortened tonic muscles and is essential for injury-free golf or any form of strenuous activity that may be hindered by muscle length imbalances. When you perform any exercise with altered muscle length and/or muscle balance, your nervous system compensates for this imbalance. The long-term result is the development of faulty movement patterns that may lead to injury during or because of exercise.

Developmental stretching is best performed at the following times:

- **Prior to exercise**
  When stretching before exercise, it is very important to only stretch the tonic muscles that your testing identified as being short. This is because developmental stretching changes the length of the muscles enough to confuse the brain's movement monitoring system and can result in unwanted joint and muscle injury.

- **At the end of the day, before going to bed**
  It is important for those needing developmental stretching to make time to stretch at night. This can be done one or two hours before bed, while watching television, or relaxing. After any form of loading that breaks the muscles down, the muscles always react by getting slightly shorter and tighter; you may have felt these tight muscles after exercise. A significant amount of healing takes place when the body is at rest. If you go to bed with muscles shortened from work, golf, etc., then your muscles will heal progressively shorter. Developmental stretching before you sleep restores muscle length and allows the muscles to heal at their optimal length.

MAINTENANCE STRETCHING

Once you have achieved balance in your musculoskeletal system through the use of developmental stretching and corrective exercise, you can begin a maintenance stretching program. Such a program focuses on problem areas and on muscles particularly stressed by golf and serves to preserve muscle balance.

Maintenance stretching should be done by everyone, not just those who feel tight. You may have found that you have adequate flexibility after performing all the tests in Chapter Two. Yet if you continue to exercise, play golf, work and so on, you will become progressively more restricted in range of motion until you can no longer achieve optimal swing mechanics. In fact, many golfers only get serious about stretching when their swing becomes erratic or their bodies begin to hurt. This frequently leads to a short-term devotion to developmental stretching, or yo-yoing between good and poor function. To prevent this from happening, maintenance stretching should be used.

PRE-EVENT STRETCHING

Many people do not naturally possess the level of flexibility needed for optimal swing mechanics. For others, such factors as sitting for long periods of time at work, old orthopedic injuries and current aches and pains tighten them up in just the span of a few hours. For this very reason, everyone should participate in pre-event stretching and golf warm-up exercises, which may include Muscle Energy Mobilization exercises, described in Chapter Four.

Pre-event stretching must be performed correctly in order to get the most from your body during golf, be it driving practice, lessons or actually playing the game. If the wrong type of stretching is performed on the wrong muscles, then joint and/or muscle injury is likely. For example, the static holding techniques used to elongate shortened muscles in the developmental and maintenance stretching phases can actually sedate the muscles being stretched. This is not desirable before exercise in muscles performing primarily joint stabilizer functions.

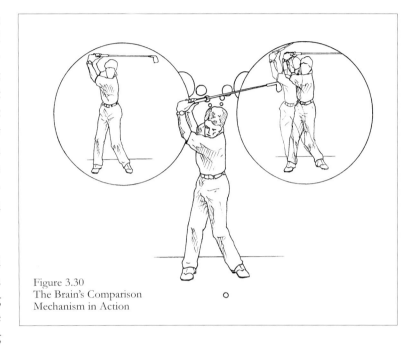

Figure 3.30
The Brain's Comparison Mechanism in Action

It is believed by scientists who study motor learning that the body stores movement patterns in the brain. Human beings have a tremendous variety of movement skills and abilities that are stored as "general movement patterns."[1,2] General movement patterns are patterns with similar qualities shared by many movements. For example, chipping, driving or pitching a golf ball all require the brain to activate the "trunk rotation" pattern, or a pattern containing the relevant timing sequence of the golf swing. From there, the brain alters the necessary variables of the chosen pattern (intensity, speed and timing) to meet the demands of the particular swing and environment.

When you perform static stretching with prolonged holding positions as traditionally prescribed to athletes as part of their warm-ups, the muscle spindle cells (length monitoring cells in muscle) are lengthened without the brain being able to effectively monitor the length change. If you go out to drive the ball after static stretching, your brain immediately notices that the information coming from the muscles used to drive the ball does not match the information stored in the brain as "trunk rotation-golf swing-driver." This results in the brain's comparator mechanism quickly trying to figure out how to modulate the swing to make it match what is stored in the brain as the "gold-standard swing" (Figure 3.30).

Unfortunately, this process often takes longer than the very short amount of time between the initiation and the impact phase of a golf swing. You may have experienced such an instance when you initiated a swing and before your club was halfway to the ball you could "feel" that there was something wrong with the swing, but you couldn't respond quickly enough to change it. If the muscles are not controlled properly by the nervous system due to prolonged stretching, the result is potential injury to muscles and joints; the nervous system cannot regulate muscle forces, control joint forces and ranges of motion fast enough to respond appropriately.

EXCEPTIONS TO THE RULE!

It should come as no surprise that there are exceptions to every rule. Because of the way our bodies are designed, there are three different types of muscles and they don't all respond to stretching the same way. The body is comprised of "mixed muscles," which simply means that the muscles contain some fibers that are classified as FAST TWITCH or fast contracting, and some SLOW TWITCH or slow contracting. Slow twitch muscles are best suited for prolonged work situations. For example, novice golfers may hold their address posture for as long as a minute before executing their shot. To hold your body in the address position for that long requires the use of postural tonic muscles, which are predominantly slow twitch.

When actually swinging the club to drive, chip, or pitch the ball, you must use both the slow twitch dominant postural muscles to hold your trunk angle and your fast twitch dominant, or phasic muscles, to accelerate your torso, arms and the club. The third group of muscles consists of muscles that are fairly equal in composition and don't fall into either the tonic or phasic group.

The interesting part is that the muscles containing a dominance of slow twitch muscle fiber, which are classified as tonic, or postural muscles, react to faulty loading situations in a very different way than the faster phasic muscles.

For example, when you go to the driving range and decide to hit an extra bucket of balls, you may overwork several of your tonic muscles and phasic muscles. If you overload the muscles enough to provoke a response, you will find that the tonic muscles react by shortening and tightening, while the phasic muscles react by lengthening and weakening. But that's not all...the tonic muscles not only shorten and tighten in response to faulty loading, they also become more and more excited and try to do the work for the related phasic muscles.

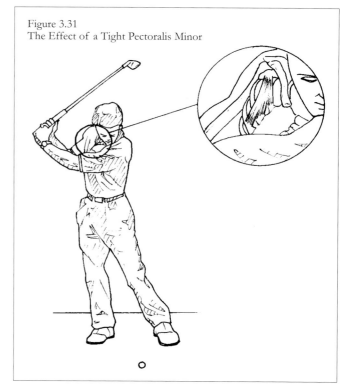

Figure 3.31
The Effect of a Tight Pectoralis Minor

This is very important information for the golfer wanting the most out of his/her game, because any shortened tonic muscles that are involved in your golf swing will rob the nerve energy from other important muscles and alter your swing mechanics. For example, if your right pectoralis minor (a tonic muscle) is short and tight, you will be unable to achieve a good backswing position (Figure 3.31). A tight left pectoralis minor will disrupt your ability to achieve an optimal follow-through.

HOW TO PERFORM PRE-EVENT STRETCHING

With the exception of shortened tonic muscles (as described above), pre-event stretching is performed without the use of prolonged static holding positions. The best way to perform pre-event stretching is to use "dynamic stretching." With dynamic stretching, you stretch your target (tight) areas without stopping, simply moving into and out of the stretch position repetitively until you loosen up.

The key things to remember are to never move fast and never stop moving for more than a second or two. When you keep moving, the brain constantly monitors the changing length of the muscles. This prevents you from experiencing joint destabilization and coordination deficits that can come from stretching with prolonged static holding techniques.

POST-EVENT STRETCHING

Stretching after any athletic event is a good means for reducing post-exercise muscle soreness. When muscles fatigue, they tighten and the blood flow through them is reduced. However after an athletic event, the metabolic rate often stays elevated for several hours with the muscles continuing to produce metabolic by-products. Many of these post-exercise metabolic by-products are acidic in nature and aggravate sensory nerve endings in muscle tissue. This serves to tighten the muscle even more. To prevent unnecessary and unwanted muscle soreness after golf, it is a good idea to perform post-event stretching.

Post-event stretching involves stretching all the major muscle groups used in golf, particularly the hips, low back and shoulders. Many golfers will not spend the time doing a thorough stretch after spending several hours on the course, but stretching the following key areas will help in preventing unwanted muscle soreness.

- Lunge Stretch – Figure 3.15
- Lumbar Erectors Stretch – Figure 3.16
- 90/90 Hip Stretch – Figure 3.17 A&B
- Indian Sit - Figure 3.20 or Groin Rocking Stretch – Figure 3.21 A&B
- Supine Knee Extension Hamstring Stretch – Figure 3.23
- Levator Scapulae Stretch – Figure 3.3
- Medial Shoulder Rotators & Pec Minor Stretch – Figure 3.5 A or B
- Lateral Shoulder Rotators Stretch – Figure 3.6
- Latissimus Dorsi Stretch– Figure 3.11
- Trunk Rotation Stretch – Figure 3.14

CHAPTER THREE

Another very important distinction to make here is that post-event stretching should never replace developmental or maintenance stretching! Stretching after golf is used to calm the nervous system, restore circulation in golf-specific muscles to optimum levels, and regain freedom of movement after the game. Your developmental and/or maintenance stretching should never be disregarded just because you performed post-event stretching or you will rapidly lose your flexibility and experience an increase in technical swing errors on the golf course!

CHAPTER 4

HOW TO WARM-UP FOR GOLF

A thorough pre-golf warm-up is something that many people ignore, being more interested in getting out on the course as quickly as possible. A golf specific warm-up will lubricate your joints, warm your muscles and connective tissues, activate your nervous system and sharpen your senses. All in all it will help improve your game!

At the C.H.E.K Institute, our athletes have achieved great results using muscle energy exercises in a warm-up routine. These exercises are derived from the well-known Feldenkrais System of Movement and modified specifically for golf by Feldenkrais practitioner Michael Rubano. Muscle energy exercises actively mobilize joints and can also be used as a treatment for movement restriction. In many instances, the golfer's body responds better to this type of stimulus rather than directly stretching muscles crossing tight joints. This is frequently the case with people above age 40, since the aging process produces degenerative changes in joints, which decreases mobility unless a deliberate effort is taken to maintain flexibility.

> *A golf specific warm-up will lubricate your joints, warm your muscles and connective tissues, activate your nervous system and sharpen your senses.*

The body often limits joint range of motion as a guarding response to protect sore joints. During stretching, muscles may actually be tightened to protect the joint from being stretched through too great a range of motion, regardless of whether that is the actual intent. Dynamic warm-up and muscle energy exercises work better than static stretching for mobilization of joints as the body feels less threatened when it has an active part in the process and allows the joints to move much more freely. The increased joint range of motion not only lasts longer in many cases, but the body learns how to move through a new range of motion without over-protecting the joints.

CHAPTER FOUR

▶18 When performing muscle energy mobilization exercises, the brain is fully aware of the new range of motion created in the joints, so the comparator function of the brain is not activated, preventing the brain from attempting to correct the movement pattern. Compare this with static stretching (Figure 3.30) where the brain is unaware of the change in muscle length created by the stretching and so it tries to compare and change movement patterns during a golf swing, resulting in poor performance.

Muscle energy mobilization exercises also serve as an excellent form of active stretching. If better results are experienced with muscle energy exercises than stretching alone, or with a combination of stretching and muscle energy exercises, simply do what works best. Progress can be checked by repeating the tests in Chapter Two.

HOW TO USE MUSCLE ENERGY MOBILIZATION EXERCISES FOR GOLF

To make the best use of Rubano's muscle energy mobilization exercises in preparation for golf, just choose the exercise(s) targeting your primary area of concern and perform it at the beginning of your dynamic warm-up exercises. This will give you maximum mobility with minimum chance of injury through your game and most importantly, these golf warm-up exercises will sharpen your nervous system and assist in developing fluidity in your swing.

With the exception of the Golfer's Neck/Trunk Trainer, the muscle energy mobilization exercises listed below require that you lie on the floor. Although this may seem impractical as a warm-up for golf, taking time to find a clean piece of carpet in the corner of the clubhouse and doing the exercises will provide valuable improvement in your golf performance! To assess which exercises provide the greatest increase in swing performance, take out a club and swing it a few times before doing each of Rubano's golf warm-ups. Indicators of improved swing performance that may be identified without actually hitting the ball are:

- Increased range of motion in the shoulders, spine, or pelvis.

- Increased fluidity of swing and a sense of reduced effort.

- Heightened activation of senses, such as sight, movement awareness, clarity of thought and even hearing may be improved!

When performing the warm-up exercises just prior to hitting practice, golfers commonly experience increased consistency, distance and a reduction in previously problematic swing faults.

If circumstances require that you warm-up in a standing position, use Rubano's Golf Neck/Trunk Trainer as the keystone of your warm-up. This exercise should be used as the last warm-up exercise performed before swinging the clubs.

Golfer's Neck/Trunk Trainer

The Golfer's Neck/Trunk Trainer is an excellent exercise for activation of the trunk, neck, shoulders and nervous system in preparation for golf. The exercise has three phases of progression.

PHASE I: Trunk Rotation Test

- Standing in a natural stance, raise your right arm so that it is reaching out in front of you, as though you were pointing at an object in the distance (Figure 4.1).

- Keep your feet firmly planted as you rotate your trunk, reaching around the body as though you were looking over your right shoulder to back your car up without using the rear view mirror (Figure 4.2).

- Pay close attention to how far your trunk rotates and how far you can reach behind your body with your arm before you are restricted by spinal rotation. Do not allow the hips to rotate.

- Repeat this test on the other side. Once again, pay special attention to how far you can reach and rotate your trunk. Now, progress to Phase II.

PHASE II

- Starting with your arm outstretched in front of you again, move your arm laterally, maintaining a horizontal plane.

- As you move your arm away from your body, hold your head still and follow your hand with your eyes.

- Do not allow your torso to rotate.

- Repeat this ten times on each side and then progress to Phase III.

Figure 4.1
Golfer's Neck/Trunk Trainer: Phase I Start

Figure 4.2
Golfer's Neck/Trunk Trainer: Phase I End

CHAPTER FOUR

PHASE III

- Performing the same movement as in Phase II, you now simultaneously rotate your head in the opposite direction while keeping the eyes on your hand as it moves away from your eyes. Initially, this may be a little challenging! (Figure 4.3)

- Do this ten times on each side and then repeat the trunk rotation test (Phase I). You should now be able to rotate your trunk, shoulder, arm and neck much further!

How did this happen? This phenomenon occurs as a result of a complex interplay between your eyes, the joints in your neck and the reflexes contained in the neck. It is very important to include Rubano's Golfer's Neck/Trunk Trainer warm-up exercise any time you want to play your best golf!

Figure 4.3
Golfer's Neck/Trunk Trainer: Phase III

Shoulder/Spine Integrator

PHASE I

- Lie on your side, hips flexed to 90°, with a towel or pillow just big enough to maintain good neck alignment placed under your head (Figure 4.4).

- Place your top hand on your forehead and gently rotate your neck as far as comfortable.

- Perform this motion slowly 10-20 times, allowing your neck to rotate a little further and your arm to drop a little closer to the floor each time. It is very important that you complete these motions using as little energy as possible. Working harder is not smarter in this case!

- After completing 10-20 repetitions, change sides and repeat.

PHASE II

- Assume the same starting position as in Phase I, but begin by placing the hands together in the outstretched position.

- Slide the top hand along the palm and forearm of the bottom arm, going only until you feel some mild restriction to rotation in your spine and/or shoulder (Figure 4.5A).

- When you feel restriction, return the top hand along the arm to the start position. **Again, it is very important not to use a lot of effort during this exercise!** Try to stay completely relaxed. The more tension you generate, the less effectively you will mobilize the joints and tune the nervous system!

- After performing 10-20 repetitions, or as many as it takes to feel as though you have loosened maximally, roll over and perform the exercise on the opposite side. You will probably find that the opposite side responds much quicker, requiring less repetitions than the first side. This is because of what is called "transfer" in the nervous system. The body always learns from performing an exercise on one side, facilitating learning on the other side. Figure 4.5B shows Phase II being performed with a greater range of movement.

- Upon completion of Phase II on each side, progress to Phase III.

Figure 4.4
Shoulder/Spine Integrator: Phase I

Figure 4.5A
Shoulder/Spine Integrator: Phase II

Figure 4.5B
Shoulder/Spine Integrator: Phase II
Greater Range of Motion

PHASE III

- Phase III is performed much the same as Phase II, with the exception being that now your top leg is moving forward over the bottom leg, the feet stay together and the arms remain still (Figure 4.6).

- Start out with very small movements, progressing to larger movements of the top leg.

- Let the torso move naturally. Do not try to keep the movement isolated to the pelvis. Again, move with the least possible amount of energy expenditure.

- Perform ten repetitions and then repeat on the other side.

Figure 4.6
Shoulder/Spine
Integrator:
Phase III

PHASE IV

- Phase IV combines Phase II and Phase III, achieving maximal benefits. You will now move the top leg forward over the bottom leg and the top arm backward across the bottom arm. This movement will now integrate the upper and lower portions of the trunk.

- Perform the movement with as little effort as possible ten times per side. The movements need not be large to be effective. Anyone watching you perform the exercise should see no signs of muscular exertion.

Hip and Pelvis Integrator

The Hip and Pelvis Integrator is particularly useful for the senior golfer or any golfer who is tight in the hips and low back.

- To begin, lie on your back and bend your left knee (Figure 4.7A), with your right arm at your side.
- To initiate the exercise, place just enough pressure on the left foot to overcome the resistance of gravity against your pelvis, barely lifting your left buttock off the ground (Figure 4.7B).
- Perform 10-20 repetitions, progressively rolling the pelvis forward, lifting just a little more of your spine off the ground each repetition.
- Be very careful to relax! With each rep, allow the hips to open up. As you progress up the spine, the chest will begin opening up provided you don't tighten up.
- When you have completed the left side, you will be easily rotating the body to the point where the shoulder is coming off the ground and you are facing forward.
- Repeat the process on the opposite side, performing as many repetitions as needed to loosen the spine and pelvis.

> 18 This is a very valuable exercise for the golfer because it not only loosens the spine, it mobilizes the pelvic girdle and sacroiliac joints. A mobile spine and pelvis are essential to good timing, rotation, coil and power generation!

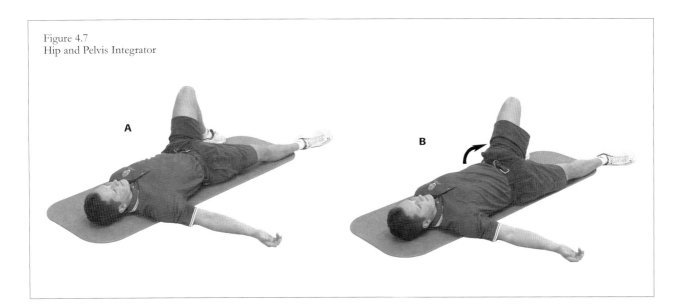

Figure 4.7
Hip and Pelvis Integrator

CHAPTER FOUR

SELECTING THE BEST MUSCLE ENERGY MOBILIZATION EXERCISE FOR YOUR WARM-UP

Although the muscle energy mobilization exercises require only a few minutes to do, they are well worth it! Pick one or two of the muscle energy exercises, along with the Golfer's Neck/Trunk Trainer, to be part of your warm-up routine.

If time is an issue, choosing between the Shoulder/Spine Integrator and Hip and Pelvis Integrator will be based on your current functional status. If you have shoulder discomfort or any recent history of shoulder dysfunction/pain, you will find the Shoulder/Spine Integrator most useful to your game performance.

Should low back pain or stiffness be a hindrance to your game, you will get the most benefit from the Hip and Pelvis Integrator. If your middle back is a problem area, you will be best served to experiment with these warm-up exercises and determine which ones give you the best results. You may find that performing them both in succession is best.

Now that you have chosen the ground-based muscle energy mobilization exercises that will best suit your needs, you can progress through the remainder of your pre-golf warm-up.

HOW TO WARM-UP FOR GOLF

PRE-GOLF WARM-UP EXERCISES

These exercises should be performed in addition to the muscle energy mobilizations.

Shoulder Clocks

The Shoulder Clock exercise is an excellent way to warm-up the biggest joint in the shoulder complex, the scapulo-thoracic joint. The scapulo-thoracic joint consists of the shoulder blade and its contact with the rib cage. This is a very important warm-up exercise for the golfer wanting to minimize any chance of shoulder discomfort due to golf.

The Shoulder Clock may be performed either standing or side lying (Figure 4.8). For the golfer who is tight in the shoulders and has not devoted any significant time to stretching, the exercise will produce better results initially from the side lying position. Once you have regained coordination and can perform the exercise fluently, you will then be able to get good results in a short period of time in the standing position.

- Lie on your side with your knees bent comfortably.

- Pretend that your shoulder is the middle of a clock (Figure 4.8A). Elevating the shoulder straight up toward your ear would represent 12:00, depressing it downward 6:00, moving it forward 9:00, and directly opposite 9:00 would be the 3:00 position. Isolate shoulder blade movement by moving the shoulder between these key positions on the clock four or five times. Alternate between 12:00 and 6:00 several times and then between 3:00 and 9:00 several times.

- Repeat on the other shoulder.

- After doing this drill on each shoulder, progress the exercise by attempting to touch each number on the clock in turn. The exercise should be performed in both the clockwise and counter clockwise directions five times on each shoulder (Figure 4.8B).

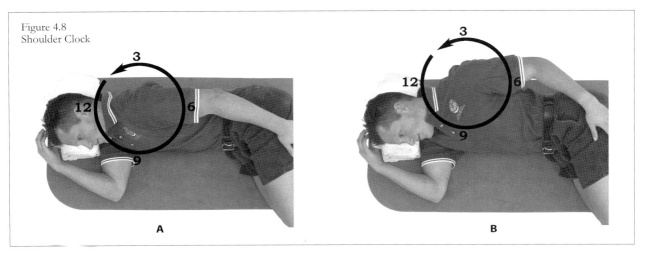

Figure 4.8
Shoulder Clock

Hip Rotations

The hips are composed of ball and socket joints, which means they can move in many directions (compared to a hinge joint such as a knee or elbow which can only bend one way).

- To warm these joints for golf, place your hands on your hips above your pockets. With your feet shoulder-width apart, begin to rotate your pelvis in circles, starting small and progressively getting larger. (See Figure 4.9.)

- Your circles should increase from small to as large as possible over the span of 12-15 hip rotations.

- Reverse the direction of movement and do another 12-15 rotations.

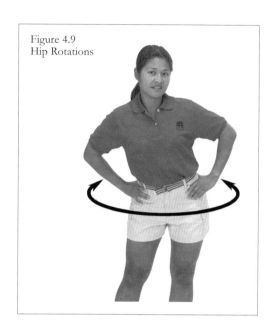

Figure 4.9
Hip Rotations

Weight Shifting

Weight shifting requires a different, more golf specific movement pattern than Hip Rotations. The Weight Shifting exercise is used to warm the sacroiliac and pubic joints of the pelvis, as their function is integral to the full swing. This exercise also warms the muscles of the pelvis, including the hip rotators and adductors.

- Place your hands on your hips at the opening of your pockets. As you shift your weight toward one leg, you should simultaneously allow the other knee to drop downward and inward slightly; this unlocks the sacroilliac joint on that side, freeing the pelvis to rotate more easily. *Care must be taken not to shift the pelvis laterally or sway, as such movements in the warm-up phase only encourage bad form on the course.*

- Start the exercise with small movements, gradually increasing the magnitude of the movement to full weight shift and pelvic rotation by about the tenth repetition on each side. When you get to the last couple of reps on each side, your weight shift should be about 80% stance leg to 20% trailing leg.

Figure 4.10
Weight Shifting

HOW TO WARM-UP FOR GOLF

▶18 The importance of allowing the trailing leg to relax lies in the function of the sacroiliac joints. These are the joints between the sides of your pelvis and the sacrum, which is the lowest segment of the spine. As you allow the trailing leg to relax, it aids the rotation of the sacrum toward the stance leg, which in turn facilitates rotation of the spine. If this function is hampered, you will be restricted in your ability to achieve optimal backswing or follow-through, depending upon which side is dysfunctional.

- To test this in your own body, place your hands together and hold them out in front of you. Stand on your left leg. Keeping the right side of your pelvis from falling down, see how far you can rotate your trunk to the left, using your hands like a pointer.

- Then take your address stance and put your hands together in front of you. Shift your weight to the left, allowing your right leg to relax, letting the knee drop toward the ground slightly and once again, rotate your trunk and arms to the left as far as you can (Figure 4.11). Surprise! You should be able to rotate your trunk at least 15-25° further to the left. If not, you probably have a sacroiliac joint dysfunction and should see an orthopedic physical therapist or C.H.E.K Practitioner for assessment.

Figure 4.11
Testing the Function of the Sacrolliac Joint

This little biomechanical demonstration should have highlighted the importance of warming up your hips and sacroiliac joints in the side-to-side movement pattern (technically referred to as the **FRONTAL PLANE** - see Chapter Two).

CHAPTER FOUR

Foot and Ankle Warm-Up

Ankle rotations have been part of the warm-up process for as long as sport has been around. Although you will not be dodging between bodies at high speeds on the golf course, you will want to warm the ankle joints in preparation for driving the ball and performing technical shots from the rough.

- Place your hands on your hips and begin to shift your weight left and right.

- As you shift left, let the arch of the left foot get big and allow your foot to roll so the weight is on the lateral edge of the left foot and the medial edge of the right foot.

- Alternate side-to-side, letting your ankles rotate and your feet transfer the weight from the medial to lateral aspect of the foot.

- Do this at least ten times on each side.

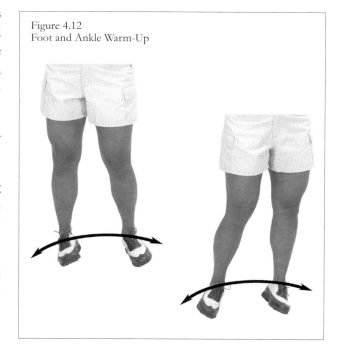

Figure 4.12
Foot and Ankle Warm-Up

🏁18 The ankle joint functions best as a hinge joint, yet the golf swing requires that the ankle actually rotate as part of a sequence of movements in related joints. Warming the ankle joints is particularly important for the golfer who may have a stiff back, hips or shoulders, as these limitations will result in excessive motion at the ankles to compensate. If your ankles are not warm, you may not only injure your ankle, but also your foot.

> ### GLOSSARY
>
> **NON-PHYSIOLOGICAL JOINT MOTION:** Think of a bicycle's steering head or the wheel bearings in your car. When you over-tighten the steering head on a bicycle or the wheel bearings on a car, the bearings get very tight and can't move. When you back the pressure off just a little bit, the bearings regain "non-physiological motion" and begin to move properly again. Similarly, a golfer's joints can't move properly if there is excess muscle tension disrupting joint motion. Non-physiological joint motion is the movement of the joint that must occur in planes of motion other than the primary movement planes of the working joint. For example, to raise the arm to the backswing position requires movement in the sagittal plane, which can't happen without freedom in the transverse and frontal planes, as well as small gliding movements.

Leg and Arm Flicks

- The Arm Flick exercise is performed by flicking your arm in a relaxed swinging motion, as though you were trying to flick something sticky off your fingers. (See Figure 4.13.)

- Leg Flicks are done by standing on one leg and flicking the free leg, as though trying to flick pebbles out of your shoe. (See Figure 4.14.)

- You should rhythmically flick each arm and leg about twenty times, progressing from mild to moderate intensity.

Leg and arm flicks relax the muscles crossing all the joints of the legs and arms by reflex activation of the nerve endings in the joint capsules (connective tissue surrounding joints). The nerve receptors in the joint capsules are sensitive to frequency of movement. *Therefore, it is very important to rhythmically flick the arms and legs or the muscles will not relax, they will tighten!* If your rhythm is poor, the receptors sense the changing rate of movement and tighten the muscles to stabilize the joints. When the movements are rhythmical, the nerve receptors adapt and the muscles relax, allowing the joints to decompress. These exercises help by lubricating the joints and by restoring what is referred to as **NON-PHYSIOLOGICAL JOINT MOTION**.

> ▶18 The whole concept of non-physiological joint motion is very important to the golfer because loss of non-physiological joint motion in any joint will lead to compensation, faulty muscle recruitment patterns and inconsistency in your swing. Generally, you will go to your teaching pro, who will do his or her best to teach you to compensate for your slice or hook, but will not realize the swing fault is a by-product of loss of non-physiological joint motion!

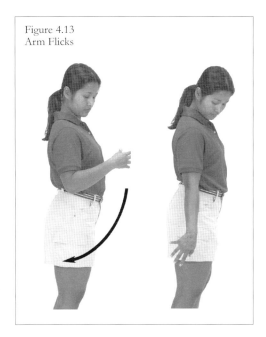

Figure 4.13
Arm Flicks

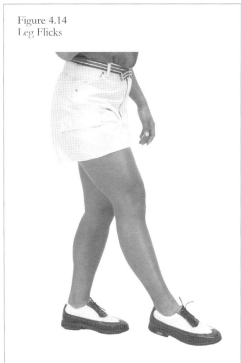

Figure 4.14
Leg Flicks

Swinging Progressions

Swinging progressions are the final stages of your warm-up for the best golf your body can give you on any given day.

- Take a seven iron and perform ten half-swings at 50% effort.

- After a short rest, perform five full swings, increasing the effort with each swing from 50% to full speed.

- For your final exercise, swing your driver ten times, progressively increasing the intensity to full swing speed and power by your tenth swing.

PUTTING TOGETHER THE PRE-GOLF WARM-UP

A pre golf warm-up should consist of four parts, in the following order:

❶ Muscle energy mobilization exercises, especially for the areas that need particular attention.
❷ Warm-up exercises.
❸ The Golfer's Neck/Trunk Trainer just prior to practice swings.
❹ Swinging Progressions.

You are now ready to play the best golf that your body is capable of playing. If you study the mental preparation techniques in *The Ultimate Game of Golf* by Bob Cisco[1], you may even play beyond what your body has to offer. Your mind is always capable of more than the body will lead it to believe!

CHAPTER 5

FUNCTIONAL EXERCISE

The Whole in One concept is built entirely upon the principles of functional exercise. Unlike exercise programs developed upon a body building format, or muscle isolation, functional exercises are designed to restore, balance, lengthen, strengthen and coordinate movement patterns specific to the individual's given work or sports environment. The human brain does not think in terms of isolated muscles. Instead, it recruits groups of muscles in uniquely programmed sequences. ***A golfer's conditioning program must therefore be designed to integrate the whole body***. Total body integration cannot be achieved through the use of machines designed to isolate muscle function. If it could, champion golfers would weigh 300 lbs. and could carry the golf cart on their shoulders!

When an athlete or worker is injured, careful consideration must be given to what is termed the ETIOLOGY (cause of injury). Today, many workers' and athletes' careers come to an abrupt halt after what should only be a temporary setback. This is because most doctors and therapists are after the quick fix. They are focused on removing pain, which is most often an indicator of a problem but not the problem itself.

When removing pain is perceived as curing a problem, the problems are just beginning. This is exactly why so many golfers and athletes of all types suffer from repeated bouts of back pain, shoulder pain, knee pain and so on. In fact, 90% of all the money spent on back pain in the U.S. is spent on the 10% of back pain sufferers who have a second, or repeat, bout of pain. Exercising regularly and correctly can keep you from becoming part of that 10% group.

You probably know someone who has had three surgeries on the same knee, or has chronic low back pain despite numerous treatments using traditional medical approaches. To truly prevent and treat injury and improve performance, a few simple principles must be followed. *Breaking these principles to get a quick fix will almost always result in short-term resolution and eventual setback.*

CHAPTER FIVE

THE RULES OF PREVENTION AND PERFORMANCE

Any corrective or performance exercise program must adhere to the following laws or guidelines if true long-term success is intended:

1. All treatment, exercise programs or otherwise, must seek to correct the underlying cause(s), or etiology, prior to attempting to improve performance.
2. Treatment should only focus on alleviation of pain during the acute phase of an injury. The acute phase generally lasts no longer than three weeks.
3. When restoration of function and/or improving performance is the goal, an exercise program must always be aimed at first restoring stability to the working joints and balance to the musculoskeletal system. Strength must then be restored. The final goal is then re-establishing and improving power output.

Virtually every book on golf conditioning breaks one or more of these principles. It is no wonder so many golfers have abandoned traditional approaches to exercise for improving golf performance.

Although it is acceptable to work on achieving optimal levels of flexibility and stability at the same time, it is critical that the golfer follows the Stability/Strength/Power format to achieve optimal golf performance as presented in this book.

THE SCIENCE OF GOLF STABILITY

What exactly does stability really mean, and how does it apply to the game of golf? There are two key types of stability, **STATIC** and **DYNAMIC**. Attempts to develop dynamic stability without first developing static stability are often futile. This chapter addresses these key biomotor abilities in the order they are best developed.

Just as you cannot expect to fire a canon from a canoe with any accuracy, you cannot expect to express the strength and power necessary for a long drive, medium iron shot, technical shot from the rough or chip shot with any accuracy if your body is unstable. You can't even putt consistently if you have the stability of a canoe. Before you can even entertain achieving your maximum range, accuracy and consistency on the golf course, you must develop the static or postural stability from which you may build strength and power.

The concept of static or postural stability is important since posture is inclusive of your physical and structural orientation. Your posture represents the physical expression of your mind's perception of posture. This is exactly why attempts to strengthen postural muscles

> **GLOSSARY**
>
> **Static Stability** is also known as postural stability. It is the ability to remain in one position for a period of time without losing good structural alignment.
>
> **Dynamic Stability** is the ability to keep each and all working joints in optimal alignment during any given movement, such that the efficiency of the movement is facilitated and injury is prevented.

without concomitant postural education are seldom fruitful. How well aligned your structure is and how well conditioned your postural muscles are can be determined in two ways:

❶ Postural Alignment
❷ Postural Sway

You Can't Fire a Canon from a Canoe!

CHAPTER FIVE

1. Postural Alignment

Postural alignment can be determined through the use of specialized instruments. Clinically, C.H.E.K Practitioners use a process of measuring alignment for the specific purpose of selecting exercises to correct postural problems and improve golf performance. A simpler approach that can be used to give a general idea of postural alignment is demonstrated in Figure 5.1.

- Hang a weighted plumb line from a door frame or other immovable object. The end of the line should fall about 1 cm. in front of the lateral malleolus (ankle bone).

- Stand in a natural position, looking straight ahead, with arms relaxed and hanging by your side.

- Much information can be gathered by noting the key landmarks of the body and their positions relative to the plumb line.

Good postural alignment (Figure 5.1A): The plumb line falls slightly anterior to (in front of) a midline through the knee, crosses the center of the greater trochanter of the femur (bony prominence of thigh bone just below the hip), falls approximately midway through the trunk and shoulder joint, through the center of the cervical bodies (neck) and finally through the lobe of the ear.

There are many different types of faulty posture; Figure 5.1B is an example of a **kyphotic-lordotic posture**. There are increased spinal curves in the lumbar region

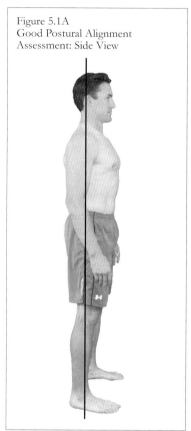

Figure 5.1A
Good Postural Alignment Assessment: Side View

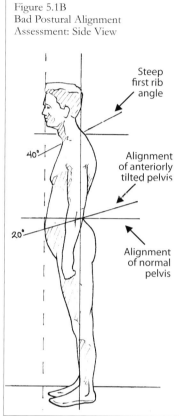

Figure 5.1B
Bad Postural Alignment Assessment: Side View

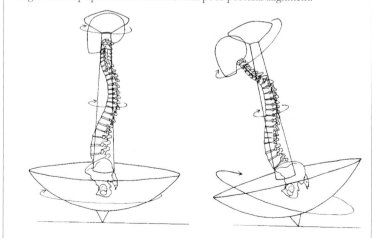

Figure 5.2
Left: The top spins true with good postural alignment.
Right: The top spins out of balance with poor postural alignment.

(lordosis) and thoracic spine (kyphosis), an anteriorly tilted pelvis, steep first rib angle and a forward head.

When a golfer possesses good postural alignment, his or her body will hold an optimal rotational axis, rotating much more efficiently. To illustrate this point, imagine your body as a top (Figure 5.2). If your postural alignment is poor, your rotational energy is expended in all directions, resulting in dissipation of energy through a constantly changing axis of rotation. This sort of spinal alignment results in poor shot consistency.

2. Postural Sway

As a golfer's postural alignment and joint stability improves, the magnitude of postural sway decreases. This happens because as the postural muscles strengthen and joint alignment improves, the body can sustain any given posture or position for longer periods of time.

A simple way to test your postural sway is to stand on two bathroom scales, one scale under each foot (Figure 5.3). Look straight forward and ask a friend to record how much variation there is in the weight on the scales. A weight shift of 0-5 lbs. is considered normal.

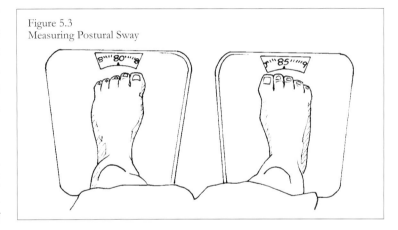

Figure 5.3
Measuring Postural Sway

This concept of improving postural stability as a means of reducing sway is a critical concept for the serious golfer to grasp. *When you have inadequate postural stability and increased postural sway, you have a very poor chance of ever reproducing a good shot consistently.*

Consider the marksman who must hit the bull's eye. If his postural stability is poor and postural sway is elevated, it is very hard to hold the rifle still and take aim. This would be like trying to make a hole in one shot from the top of one 100-story skyscraper to another on a windy day (Figure 5.4). There is no doubt that even Tiger Woods could increase his chances of putting the ball in the cup if the building were to stop swaying.

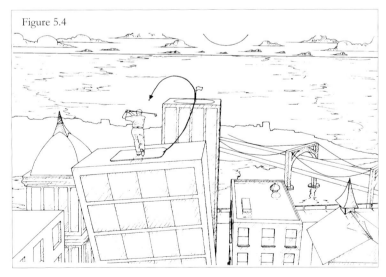

Figure 5.4

Reducing postural sway increases your chance of reproducing an optimal club path. This is because postural strength and stability create a more perfect and reliable axis of rotation, allowing more precise integration of the legs and arms with the trunk. You have probably seen a golfer leading a tournament fall apart in the last half or last quarter of the game. This is exactly what will happen to your consistency if your postural muscles can no longer stabilize your spine and provide a reliable rotational axis from which to integrate the arms and legs during the swing.

Another very important fact seldom realized by golfers is that as your postural stability improves, your technical coaching has a better chance of making a difference in your swing.

STATIC STABILIZATION EXERCISES

To improve, or minimize postural sway, you must select exercises that target postural muscles and challenge the balance mechanisms of the body. You must also work from **ISOLATION** to **INTEGRATION**. Isolation in this case means isolating a muscle or muscle group to re-establish neurological communication between the nervous system and the muscle. This must be accomplished before progressing to the Dynamic Stabilization Exercise phase, which requires a much higher level of neuromuscular integration.

Prior to attempting any of the Static Stability Exercises presented in this phase of your conditioning program, it is very important that you understand the importance of performing all the exercises with **PERFECT POSTURAL ALIGNMENT!** Your nervous system is like a computer. If you program it with poor quality information, that is exactly what will come out of it on the golf course. By performing the exercises exactly as seen in the diagrams, you will be teaching your body to function from a position of optimal joint alignment. This will provide a base of movement skill from which your brain will draw information as you progress through your program and begin integrating your new body into your golf game.

The exercise programs to develop static and dynamic stability and to progress from isolation to integration are divided into three phases:

> Phase I – Neuromuscular Isolation Exercises
> Phase II – Neuromuscular Integration Exercises
> Phase III – Dynamic Stability Exercises

The remainder of Chapter Five contains step-by-step explanations of the exercises in each of these phases. At the end of each phase, there are three programs showing how to arrange the exercises into a weekly workout.

Program Option A allows you to perform one exercise each day of the week. This is an excellent choice if you can devote 5-10 minutes each day to exercise, but find it difficult to schedule any more time than this for conditioning. Perform this option for six weeks.

FUNCTIONAL EXERCISE

Program Option B divides the exercises into a four-day per week program. Each day will take approximately 15-20 minutes. Perform this option for four to six weeks.

Program Option C is also set up for four training days per week, but in this case you perform each exercise twice. This is the ideal program, as you will make the most rapid gains. Try this program if you can make 30-40 minutes of time available for conditioning four days a week. Perform this option for four weeks.

SENIOR GOLFER ALERT!

The senior golfer should pay particular attention to the Flexibility, Static and Dynamic Stability phases of this Whole In One golf conditioning program. It is during these phases that seniors are most likely to see a dramatic improvement in their golf games. The dramatic changes will result from the corrections made to your posture and balance.

Balance tends to decrease with age since proprioceptive nerve endings (motion sensing) are lost from the spinal column over time. There is also a progressive loss of muscle mass, with the abdominal and buttock musculature tending to weaken first.[1]

The process of aging goes hand in hand with degenerative joint changes, desiccation of spinal discs and reduced range of motion. The result is usually seen as a loss of distance on long shots. Reduced joint range of motion and less than optimal postural muscle function and strength are the source of inconsistency on the golf course. When the postural muscles fatigue easily, it is very hard to hold your address angle. The inevitable result is loss of your swing axis and swing faults.

THE GOOD NEWS IS…sticking to the Whole in One flexibility and stability conditioning programs can significantly slow the natural processes of aging. It is completely normal for senior golfers to report improved balance, flexibility and golf performance. The Strength Training phase of the Whole In One program (Chapter Six) will assist the senior golfer in maintaining and even building muscle mass – muscle in the exact areas needed to improve your golf score.

CHAPTER FIVE

FUNCTIONAL EXERCISE – PHASE I
Neuromuscular Isolation Exercises: Building A Foundation for Golf Performance

The exercises in this phase of your program target muscles that are commonly weakened in today's environment of limited movement variety. These exercises are critical to re-establishing good communication between the brain and key postural and stabilizer muscles. Before progressing to more integrated exercises, it is important to ensure that the brain can communicate with these muscles. If not, progressing to the next phase will only serve to magnify muscle imbalances. Several of the exercises presented here are used to rehabilitate and prevent spinal injury of all types. Following the programs presented in this book can both improve golf performance and prevent injury.

Prior to attempting any exercise or stage of an exercise program presented in this book, you must carefully read the instructions given for each exercise. It is also very important to remember that you must stretch all tight tonic muscles prior to beginning any exercise session for optimal results (see Chapter Three). Choose the programming option that best suits your time schedule and/or time commitment to your golf conditioning program. It takes years of practice to realize that sometimes less is more. More specifically, if you try to rush the exercises thinking that more is better, you will only get more poor quality training!

There is no short-cut to true conditioning. The principles underlying the following conditioning programs have been successfully applied to many of the world's greatest athletes. If you truly want to play the best golf of your life, commit to conditioning *you*, the golfer. Remember, no club can play the game for you.

Phase I Equipment Needs

To perform your Phase I conditioning program, you will need the following items:

- A wooden dowel rod, six feet long and $1^{1/4}$ to $1^{3/8}$ inches in diameter. This is the size of a wooden closet rod. They can be purchased at most hardware stores. A PVC pipe of the same approximate dimensions will also work.

- A 45cm Swiss ball or small, soft inflatable ball, 1 to 1½ foot in diameter, such as a Gertie ball.*

- A piece of stretch band of light to medium density. The Physio Toner is recommended.*

- A blood pressure cuff with an extender hose, or a similar biofeedback device.*

* See Resources section on page 219.

HOW TO READ YOUR CONDITIONING PROGRAM

Exercise	Rest	Intensity	Reps	Tempo	Sets
Prone Cobra	<1:00		1-8	30/15 secs	1-2

- **EXERCISE:** You will see the names of each exercise in the order that they are to be performed. There may be a number following some of the exercises. This corresponds to a note at the bottom of the page. These extra instructions are often very important, so make sure you read them.

- **REST:** This column contains the information needed to regulate your rest periods. In the example above the prescription is to rest less than (<) one minute between each set. When you see this type of rest prescription, you are being directed to begin your next set as soon as you feel recovered enough to complete it with good form, but you are not to rest longer than one minute.

  Arrows in the rest column, as seen in Programming Option B (page 141), indicate that the exercises are to be performed in a circuit format. The down arrow indicates that you should progress to the next exercise, the arrows pointing right and up indicate that you return to the first exercise. Often there will be a rest time in the top box, which means that you should rest for that amount of time after completing each full circuit. Some programs contain mini-circuits or supersets. Complete them by following the arrows and rest periods indicated.

- **INTENSITY:** With regard to resistance training, "intensity" is related to a percentage of how much you could lift with a maximum effort. This column will indicate how hard you will work. Because some of the exercises require the use of a blood pressure cuff, you will see prescriptions such as + 30 mm Hg and - 10 mm Hg, which are indicators of how much the needle should rise or fall when performing that exercise. If there is nothing in the intensity column, the repetitions will determine the load. In some cases there will be a number with a minus sign, e.g. -2. This signifies that you should stop the exercise when you feel that you could still perform that number of repetitions with good form.

- **REPS:** The number of repetitions to be completed is outlined here. There is often a range of numbers, such as 8-12. In this case, you should choose a load or adjust the effort applied to allow completion of a minimum of eight reps, and not more than twelve. When you can complete more than 12 reps, you need to increase the intensity. When you are unable to complete eight reps, you must decrease the intensity. These modifications allow you to stay within what is called the "repetition window." The repetition window is used to control intensity and training volume.

- **TEMPO:** The pace at which you move during the exercise, or in some cases, how long to hold a position and how long to rest between repetitions, is the tempo. In the example above, the prescription is to hold the Prone Cobra position for thirty seconds, while relaxing for fifteen seconds between repetitions (30/15 sec). A 10/10 tempo indicates a ten second hold followed by a ten second rest.

CHAPTER FIVE

When you see the term "Max" in the tempo column, as indicated for Swiss Ball Neck Training, you are to hold the position for as long as it takes to fatigue the muscles used in any given pattern or position in which you are applying pressure to the ball.

A number prescription, such as 333, is read in the following manner. The first three indicates how long it should take to perform the first movement of the exercise. The second 3 indicates the duration of the hold between the first and second movement of the exercise. For example, during the Swiss Ball Reverse Hyper-extension, you will raise the legs for the count of three, hold the legs elevated for the count of three, and lower the legs for the count of three (333 tempo). Just remember, the tempo prescription is written in the same order that the exercise is executed.

To assist you in performing the tempo correctly, just count to yourself as you complete each motion of the exercise. For example, during a 333 tempo, you would initiate the first movement as "one - two - three," then "hold one - two - three" and "lower one - two - three." Each number count represents a second. It is very important that you follow the tempo prescriptions exactly, as the tempo at which you perform an exercise has a great deal of influence on the outcome.

When you see the word "Slow" or the letters S, M or F, this indicates Slow, Moderate or Fast. These are tempo progressions. Once you have mastered the exercise at a slow tempo, which is a 303, you should progress to a moderate tempo, which is a 202. Finally, when you have mastered the exercise at a moderate tempo of 202, you then perform the exercise at a fast tempo of 101.

- **SETS:** Here you will see how many sets you should perform. It is common to see a smaller number followed by a larger number. In the example above, the Prone Cobra is to be performed for 1-2 sets. This means that you always start the program by performing only the lowest number of sets (one set in this case). When you next do the program, if you are not sore from the last training session, add an additional set. This progression is always used, regardless of how many sets will be performed.

PHASE I EXERCISES

Exercise	Rest	Intensity	Reps	Tempo	Sets
Prone Cobra	<1:00		1-8	30/15 secs	1-2

The Prone Cobra is an excellent exercise for conditioning the postural muscles and re-establishing optimal postural alignment.

- Lie face down on a mat or comfortable surface and rest your arms at your sides.

- Elevate your torso while simultaneously squeezing your shoulder blades together and externally rotating your arms. When you have reached the proper end position (Figure 5.5), your palms should face away from your body, your head and neck should be well aligned and your toes should be touching the ground.

- Do not allow your head to roll backward. If this happens, you will perpetuate shortness of the muscles at the base of your skull and the exercise will only serve to maintain poor posture.

If you experience low back discomfort beyond the expected muscle fatigue discomfort, begin the exercise by tightening your glutes. This will reduce your lumbar curvature, encouraging the use of back muscles higher up the spine. It is also helpful in this case to stretch your back muscles just prior to beginning the exercise. (See Figure 3.16, the Lumbar Erector Stretch in Chapter Three.)

As indicated above, the exercise tempo is 30/15 seconds. This means that you will hold the Prone Cobra position for thirty seconds followed by a fifteen second recovery. This is to be done repeatedly for as many as eight times to complete the first set. Do not progress to a second set until you have performed one set and allowed your body to adapt between exercise sessions. If you don't feel discomfort from your first workout, attempt a second set on your second workout after resting for no more than one minute.

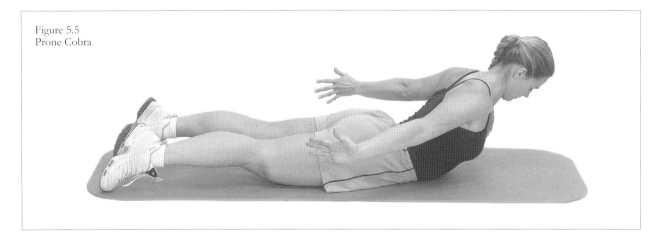

Figure 5.5
Prone Cobra

CHAPTER FIVE

Exercise	Rest	Intensity	Reps	Tempo	Sets
Horse Stance Vertical/Horizontal	1:00		10 each	10/10	1-3

The Horse Stance Vertical (HSV) and Horse Stance Horizontal (HSH) are excellent exercises for improving control and strength of many key stabilizer muscle groups. When you are first starting your program, just do the HSV.

Horse Stance Vertical

- Place your wrists directly below the shoulders and your knees directly below their respective hip joint.

- The legs are parallel and the elbows should remain turned back toward the thighs with the fingers directed forward.

- Place a dowel rod along your spine and hold perfect spinal alignment. The rod should be parallel to the floor. The space between your lower back and the rod should be about the thickness of your hand.

Figure 5.6
Horse Stance Vertical

- Draw the navel inward toward the spine just enough to create a space between your belt and your stomach.

- It is advisable to find a spotter who can assist you with feedback about your body position. If you are not training with a spotter, it is highly recommended that you train in front of a mirror to make sure you stay in correct position throughout the exercise. When you are checking your body position in the mirror, do not move your head, just look up with your eyes.

- The HSV is initiated by lifting one hand off the floor just enough to slide a sheet of paper between the hand and the floor or mat. The opposite knee is then elevated off the floor to the same height. Keep the dowel rod level at all times. Hold this position for ten seconds. After ten seconds, alternate hands and knees, again lifting them only enough to slide a sheet of paper between the extremity and the mat.

FUNCTIONAL EXERCISE

The target number of repetitions is ten per side with a ten second hold in each position. When you are able to complete the HSV for three sets with a one minute rest between sets, you are ready to add the HSH to your program. Perform one set of the HSV as a warm-up for the HSH.

The Horse Stance Horizontal

- The start position is identical for all Horse Stance exercises.

- Raise one arm to a point 45° off the midline of the body and hold it in the same horizontal plane as the back (Figures 5.7 and 5.8).

- Elevate the leg opposite the arm you have raised (left arm / right leg and vice versa) to the point at which your leg is in the same horizontal plane as your torso. As you elevate the leg, do not tilt your pelvis forward; you will know if this happens because the space between the stick and your lower back will increase. Hold the leg out straight, activating the muscles of the buttocks.

- At no point during the exercise should your shoulder girdle or pelvis lose their horizontal relationship with the floor. It is quite common for the shoulder to drop on the elevated arm side and for the hip to raise on the side of the extended leg. Either of these faults constitutes poor form!

- The arm and opposite leg are now held in this position for ten seconds before switching sides. Repeat ten times per side, providing you can maintain perfect form. Again, watch yourself in the mirror intermittently or have a spotter check your form. It is critical that you only perform as many repetitions as possible with perfect form! Failure to follow these instructions will result in futile attempts at conditioning and no improvement in golf performance. Lack of attention to detail is exactly why many exercise programs fail!

You will most likely find the Horse Stance exercises very challenging. Don't worry, you're not alone. These exercises have been used by elite athletes in numerous sports. Stabilizer weakness is a common denominator in poor athletic performance and orthopedic injury. So stay at it, do them perfectly and you will be well on your way to graduating into Phase II of your Whole in One Golf Conditioning program!

Figure 5.7
Horse Stance Horizontal: Side View

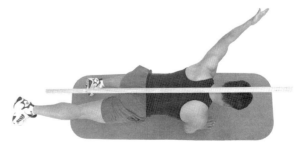

Figure 5.8
Horse Stance Horizontal: Overhead View

CHAPTER FIVE

Exercise	Rest	Intensity	Reps	Tempo	Sets
Swiss Ball Neck Training	1:00	<40%max	1-3 each	Hold 30 secs	1-3

The neck is often overlooked as a key link in golf performance. The importance of this vital link exists in the fact that the head carries instruments vital to golf success, such as the eyes and the balance mechanism in the inner ear. The point at which the head attaches to the neck, the upper cervical spine, is well known to provide a great majority of the information needed by the brain to sense movement. This movement sense is called **PROPRIOCEPTION**.

The neck is also highly integrated with the shoulder. In fact, there are several muscles running from the head and neck into the shoulder and shoulder girdle. Because of this vital interplay between the head/neck/shoulder complex, it is essential to golf performance that the neck be adequately conditioned to handle the high speed rotational movements that happen between the head and body during the backswing, downswing and impact phases. During the follow-through, the neck must rotate quickly to allow the eyes to track the flight of the ball.

The Swiss Ball Neck Training (SB Neck Training) exercises allow the golfer to begin strengthening the stabilizer functions of the neck. Light forces will be used with this exercise as stabilizer muscles effectively stabilize joints at intensity levels below 40% of maximum voluntary contraction force. Stabilization of joints commonly takes place over extended periods of time; therefore, if the muscles performing this were to require too much effort, fatigue would occur very quickly and joints would be damaged.

You will need to use a 45cm Swiss ball or similar inflatable ball about 1 to $1^{1/2}$ foot in diameter for these exercises. The ball should be inflated to the point that a 2" dent can be made in the ball when you apply firm pressure with a finger. After each use, you may want to cover the ball and store it off the floor and away from children to keep it clean.

GLOSSARY

Proprioception is the awareness of posture, movement and changes in equilibrium and the knowledge of position, weight and the resistance of objects in relation to the body.

FUNCTIONAL EXERCISE

Swiss Ball Neck Side Flexion

Your neck stabilization exercises begin with side flexion.

- As seen in Figure 5.9, the ball is placed in a doorway just high enough to clear the shoulder. The apex of the ball should rest at ear level.

- Stand with perfect postural alignment (cheek bone over collar bone and knee, hip and shoulder vertically aligned) and grasp the edge of the door with the hand on the side of the ball.

- Pull your body into the ball with a light force that you feel you could maintain for a minimum of thirty seconds. You may push into the ball with the other hand if it is more comfortable.

- Hold pressure on the ball from a position of perfect postural alignment for about thirty seconds.

- Switch to the opposite side and repeat the process.

Figure 5.9
Swiss Ball Neck Side Flexion

Perform only one thirty second hold per side. Upon beginning your next neck training session, if there is any residual muscle soreness, continue with only one repetition. If you feel no soreness, then you are ready to add another repetition. The same criteria is used to begin the third repetition. This process will apply to the following SB Neck Training exercises as well.

It is very important not to use too much force. If you can manage thirty seconds and don't feel tired, then you can increase the load on the following training session or repetition. You should now proceed to the neck rotation exercise below.

Swiss Ball Neck Rotation

- The ball is positioned much the same as it was for the SB Neck Side Flexion exercise, the difference being that now your face is positioned slightly behind the apex of the ball (Figure 5.10).

- Turn your head into the ball until you are looking straight forward.

- Hold this position for thirty seconds.

- After completion of one hold, switch to the opposite side and perform the same.

Again, if you can't hold for thirty seconds, you are pressing too hard, which will not serve to improve the stabilizer muscle functions. It will only serve to make you look like a football player and you will have to buy new shirts to fit your Hulk Hogan neck!

CHAPTER FIVE

Use the same progression plan as for the SB Neck Side Flexion exercise. You can now progress to the SB Neck Extension and Flexion exercises.

Swiss Ball Neck Extension

- Place the ball so that it is supported comfortably on the back of your head (Figure 5.11).

- Using one or both arms, push yourself into the ball. Push only hard enough to sustain the pressure for approximately thirty seconds.

Swiss Ball Neck Flexion

- From a position of perfect postural alignment, place the ball such that the ball is supported by your forehead (Figure 5.12).

- Place your tongue on the roof of your mouth just behind the front teeth; placing the tongue here ensures proper activation of the neck flexor muscles.

- With one or both hands, grasp the trim around the doorway and pull yourself into the ball. **Don't lose your alignment!**

- Pull yourself into the ball only enough to fatigue the neck flexors in about thirty seconds.

Figure 5.10
Swiss Ball Neck Rotation

Figure 5.11
Swiss Ball Neck Extension

Figure 5.12
Swiss Ball Neck Flexion

FUNCTIONAL EXERCISE

Exercise	Rest	Intensity	Reps	Tempo	Sets
Four Point Transversus Abdominis Trainer	<1:00		10	10/10	1-3

The Four Point Transversus Abdominis Trainer (TVA Trainer) is a key exercise for every golfer. The TVA is probably the most important stabilizer muscle in the body. It not only has intimate communication with other stabilizing structures throughout the body, but its activation has also been shown to precede arm or leg movement.[2,3]

Due to the extremely high compressive forces and torque placed on the spine by the golf swing, lack of TVA activation will certainly be a catalyst for both reduced swing performance and increased likelihood of injury![4]

- Assume the start position shown in Figure 5.13.

- With the spine in neutral alignment, take a deep breath in and allow your belly to drop toward the floor.

- Exhale and draw your navel in toward your spine as far as you can as shown by the arrow in the picture. Once the air is completely expelled, hold the navel toward your spine for ten seconds, or as long as you comfortably can without taking a breath (not longer than ten seconds).

- Breathe in and relax your belly. Repeat the exercise for a total of 10 reps per set. Throughout, keep your spine and back motionless.

After completing a set of the TVA Trainer, follow your program instructions for the rest period. As you are able, build up to completing three sets of the exercise. In later phases of your conditioning program, the TVA Trainer will become more complex to match the progress of your nervous system.

Figure 5.13
Four Point Transversus Abdominis Trainer

CHAPTER FIVE

Exercise	Rest	Intensity	Reps	Tempo	Sets
Greg Johnson's Window Washer	L/R		12-20 each	Slow	1-3 each

This excellent exercise for strengthening the shoulder girdle musculature was developed by physical therapist Greg Johnson. When you consider that almost 50% of all golf injuries are arm related, it becomes evident that the golfer must have a well-coordinated and well-conditioned shoulder/arm complex.[5]

The muscles that attach the shoulder blades to the spine and ribs play a vital role in protecting the shoulder and arm from injury. It is the job of these muscles to constantly position the shoulder socket in the right place at the right time to ensure optimal shoulder and arm function. If the muscles around the shoulder blade become weak, fatigued or poorly coordinated, then arm injuries are a likely result for any golfer.

To perform Greg Johnson's Window Washer, you will need a device called the Physio-Toner (see Resources section) or a piece of light stretch band, stretch cord, or even a piece of $^3/_8$ inch surgical tubing from the drug store.

Figure 5.14
Greg Johnson's Window Washer

- Grasp the Physio-Toner in your right hand.

- Place the handle of your Physio-Toner under your left thumb.

- Lift your left arm up, as if you were going to clean a window.

- Maintain good upright posture.

- Make clockwise circular motions with your left arm and shoulder. To get the most from the exercise, try to use your arm as little as possible by making the circular cleaning motions come from your shoulder blade; the arm and shoulder blade should move as a unit.

- Perform 12-20 repetitions and switch hands. If you are able to do more than twenty repetitions before feeling fatigued, you need to place more tension on the elastic or progress to the next level of band resistance.

- Change the direction of the arm motion from clockwise to counterclockwise during each set.

FUNCTIONAL EXERCISE

Exercise	Rest	Intensity	Reps	Tempo	Sets
Swiss Ball Hyper-Extension	1:00		8-12	333	1-3

The Swiss Ball Hyper-Extension is an excellent exercise for isolating the low back, gluteus maximus and hamstring muscles. Unlike the machines in the gym that isolate each muscle individually, the SB Hyper-Extension not only strengthens the muscles, but trains them to work more effectively together (Figure 5.15A). This is a particularly important exercise for the golfer because these muscles act to stabilize your torso at address as well as assist in driving your trunk and arms through the swing.

To get started, you will need a secure place to anchor your feet. If you can, it is a good idea to have a partner hold your feet as shown in the diagram.

- Beginners should start with the apex of the ball placed in the region of the upper third of the thigh. As you get stronger, the exercise can be made more challenging by moving the ball toward the knees.

- Place your arms at your sides. When able to perform three sets of the exercise at the prescribed number of repetitions, you may then place your arms across your chest as seen in Figure 5.15A. The next progression is to place your finger tips behind your ears. Raising the weight of your arms upward makes the exercise more challenging.

- Drop down over the ball until you approach the floor or until your hamstrings become stretched. It is important to try and keep your lower back from rounding forward; you need to try to maintain your natural lumbar curvature.

Figure 5.15A
Swiss Ball Hyper Extension

Figure 5.15B
Swiss Ball Hyper-Extension: Advanced

CHAPTER FIVE

- When you come up from the bottom position, make sure you stick to the tempo. It is also very important to keep the head in line with the shoulder, torso and hip. You should never look slouchy or out of alignment above the waist.

- When you can perform three sets of 8-12 repetitions using the prescribed 333 tempo with your finger tips behind your ears, you are ready to try the advanced version (Figure 5.15B).

The advanced version requires coming to a more vertical position. This is much more demanding of your hamstring muscles. To perform the advanced version of the SB Hyper Extension, always warm-up with a set of the easier version to prepare the muscles. *It is very important not to attempt the advanced version until you are able to complete three sets of 8-12 reps of the easier version with fingertips behind ears. If you perform the advanced version before your body is ready, you could easily tear a hamstring muscle!*

Exercise	Rest	Intensity	Reps	Tempo	Sets
Lower Abdominal #1-3	<1:00	+30 mmHg	**	S/M/F	1-3 each

**See directions on pages 110-112 for the repetition range for each progression.

The lower abdominal muscles play a key role in stabilizing not only your spine, but your entire musculoskeletal system during the golf swing![6-10] The abdominal muscles are also key muscles to condition, as research shows that they tend to weaken with age.

Surprisingly, there are many so-called experts on conditioning writing articles, lecturing and producing videos stating that there is no such thing as lower abdominals. This is far from the truth! Joel E. Goldthwait et.al., clearly defined the differentiation of lower abdominal muscles in the book *Essentials of Body Mechanics In Health and Disease*, published in 1934.[11] He also demonstrated several useful lower abdominal exercises.

Today, with the help of current information from Australia, these exercises have been refined, making them much more effective for the golfer.[12-14] To begin, you will need a blood pressure cuff (BPC) with an extender hose. This unit serves as a pressure biofeedback unit. (See Resources section for options.)

The BPC is a very useful tool for preventing overuse of the larger rectus abdominis muscle, as well as preventing golfers from flattening their backs too much. These are important points because the rectus abdominis is not designed to be, nor is it best utilized as, a primary postural stabilizer. Also, flat backs are becoming common among golfers because the seated workplace is now the number one workplace in the world, encouraging flattening of the low back and rapid disc degeneration.[15]

Lower Abdominal #1

- Lying on your back with your knees bent and feet flat on the floor, slide your hand under your low back at the level of your belly button. Once the thickest part of your hand is under your spine, get acquainted with the feeling you have with your spine arched to the degree it is when supported by your hand.

- Remove your hand and quickly place the BPC under your low back directly behind your navel (Figure 5.16). Pump the BPC up until you feel the same relative arch being created as you felt with your hand under your low back. Once you've identified the pressure that restores your lumbar curve to the same degree your hand did, you will add 30mmHg to that in order to achieve "your" optimal starting pressure. This is generally around 40mmHg, but you may need more or less air in the BPC.

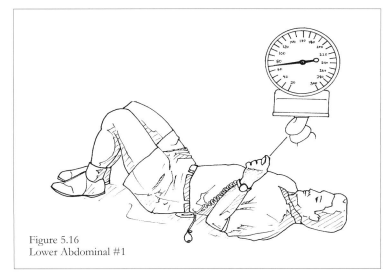

Figure 5.16
Lower Abdominal #1

- To begin, exhale and draw your navel inward toward the spine as you gently tilt the pelvis posteriorly. The BPC will now be compressed and the gauge should read level of pressure 30mmHg higher than your start pressure. For example, if you start with the BPC gauge reading 40mmHg, you will increase the pressure to 70mmHg.

- Your goal is to relax the entire body (jaw, neck, shoulders, trunk, and legs) and hold the needle at +30mmHg for ten seconds. After ten seconds have passed, or as long as you can hold the needle at +30mmHg, you should rest for ten seconds.

- Repeat this process ten times to complete one set.

Initially, you may find it hard to do the exercise and not hold your breath. It is very important to practice the exercise as often as possible, making every attempt to hold the pressure on target for ten seconds without tensing up or interrupting your natural breathing pattern. If you are unable to do this, you should seek the help of a C.H.E.K Practitioner (see Resources section) or a physiotherapist trained in corrective exercise for orthopedics.

As you become more proficient at performing the exercise, you can make it more challenging by placing the feet progressively further from your buttocks. When you can perform ten repetitions on the 10/10 tempo, you are ready to graduate to Lower Abdominal #2.

Lower Abdominal #2

Lower abdominal #2 (Figure 5.17) is performed in much the same manner as Lower Abdominal #1. The placement of the BPC is identical. The difference is that once you get the cuff pressurized to +30mmHg above your startng point, you slowly lift one foot off the ground, always trying to keep the BPC needle within a 10mmHg window. For example, if you start at 40mmHg and increase the pressure to 70mmHg by gently drawing in your navel and posteriorly tilting the pelvis, then the gauge should not move above 75mmHg or below 65mmHg as you lift and lower each leg. Optimally, elevate the foot so that the thigh is perpendicular to the floor. If you find you have a hard time maintaining the pressure within the 10mmHg working range, try moving the leg in smaller increments.

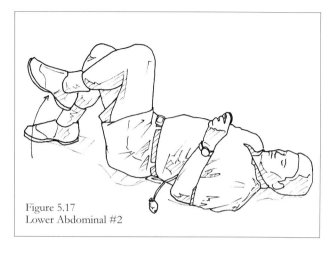

Figure 5.17
Lower Abdominal #2

It is important to always alternate between your left and right legs, as that is how you walk. When you can do 12-20 repetitions on each leg with perfect form, you can increase the challenge by slightly straightening each leg. This increases the length of the lever arm working against the lower abdominals, making it harder for them to stabilize the pelvis. When you can perform the prescribed repetitions with an almost fully straightened leg, you are ready to begin to manipulate the tempo. This is done by progressing from the slow pace of movement you started with (303) to a moderate pace (202), and then finally to a fast pace (101). Having mastered keeping the BPC needle in the target window with a nearly straightened leg at various speeds of leg movement, you are now ready to progress to Lower Abdominal #2-B.

Lower Abdominal #2-B

This exercise is quite a big step forward in complexity, although it may not look like it! Start with both feet off the ground. The support previously provided by the hamstring and butt muscles of the leg that was on the floor will no longer be present. Now you must rely completely on your lower abdominal muscles to stabilize your pelvis.

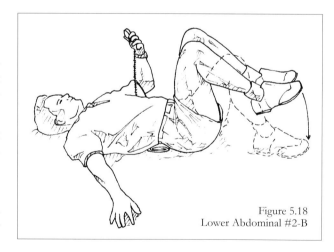

Figure 5.18
Lower Abdominal #2-B

Alternate between lowering and lifting the right, then left leg. Progress the exercise exactly as you did in Lower Abdominal #2. When you can perform the exercise with either leg straightened, in an alternating pattern, for the prescribed number of repetitions (see pages 110-112), you will be ready for your next challenge, Lower Abdominal #2-B Standing.

Lower Abdominal #2-B Standing

Lower Abdominal #2-B Standing is designed to integrate your new lower abdominal strength and coordination into a functional standing position. This is a very important transition to make because your brain must learn to activate these muscles in the position from which you play golf.

- Stand in the doorway or by a vertical post as demonstrated in Figure 5.19.

- Place the cuff directly behind the navel. Pressurize the cuff the same as in Lower Abdominal #1 and 2.

- Move your leg backward and forward in much the same manner as when walking, keeping the pressure gauge in a 10mmHg working range. You will need to concentrate on not tightening the body. It is very important to keep your head on the wall or post behind you so you don't slouch into poor posture. Your brain constantly monitors your posture as you exercise, and the last thing you want is the brain thinking that poor posture is normal!

While performing the exercise be very aware of your breathing. When your breathing is strained, it disrupts your normal respiratory cycle. Keep the amplitude of the movements within your capacity to stabilize, yet not straining to the point of disrupting diaphragm movement. You should not feel like you are interrupting your breathing to stabilize.

The exercise is progressed by speeding the tempo of leg movement as you get better at holding the needle in the target zone. When you feel confident, you can try performing the exercise freestanding without the BPC. When doing the exercise freestanding, place one finger in your navel and one finger on your spine directly behind the finger in your navel. Your goal will now be to try to move either leg without letting the spine move. The same repetitions, tempo and other exercise variables apply to this advanced exercise.

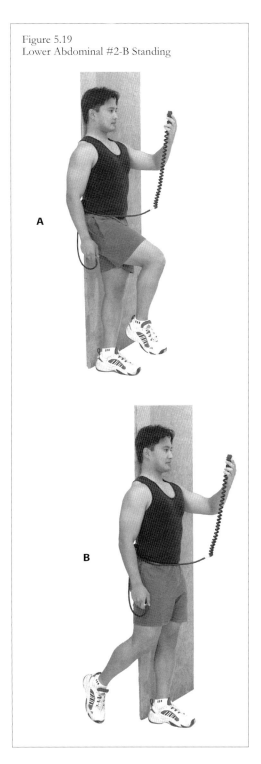

Figure 5.19
Lower Abdominal #2-B Standing

CHAPTER FIVE

Lower Abdominal #3

Lower Abdominal #3 is a progression on Lower Abdominal #2 and #2-B in that it allows greater development of strength by using both legs to load the lower abdominals (Figure 5.20).

This exercise is performed exactly the same as Lower Abdominal #2, but with two legs moving together. As you get stronger, straighten your legs. When you can keep the BPC needle in the target zone and lower your legs to the floor and back up again, you will have completed your basic lower abdominal training. You will now only need to do the exercise every week or two just to remind the nervous system how to do the exercise. The other exercises in your later conditioning phases will help keep your lower abdominals strong as well.

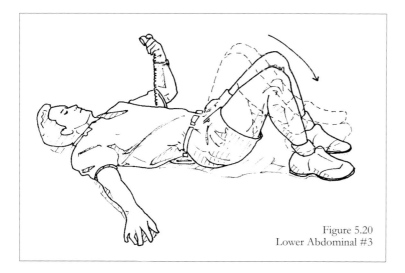

Figure 5.20
Lower Abdominal #3

IMPLEMENTING PHASE I

Below are three scheduling options by which you can carry out your Phase I – Neuromuscular Isolation Exercises. The programs are designed to last four to six weeks. If you don't feel you have mastered the exercises or have had schedule interruptions, then six weeks is fine. If all has gone well and you have been consistent, progress to Phase II – Neuromuscular Integration Exercises after four weeks.

If there are exercises you don't understand in any phase of your Whole In One Golf Conditioning program, or that cause discomfort, you should consult a C.H.E.K Practitioner or CHEK Golf Biomechanic or seek help from a qualified physical therapist trained in the C.H.E.K system. Most of the exercises in this book are demonstrated on DVDs available from the C.H.E.K Institute. You may find it very valuable to review the DVDs containing the exercises of concern to you. (See the Resources section).

In Phases I-III, the exercises are periodized in a linear format. Linear periodization is the progressive increase in loading by volume or intensity. The exercises are neuromuscular, meaning that they are intended to increase the communication between the nervous system and the specific muscle(s) being targeted. With neuromuscular exercises, the intensities are seldom high enough to be concerned about breakdown from overloading, therefore linear periodization works well.

CARDIOVASCULAR EXERCISE

Cardiovascular exercise is particularly important for the health of your heart. Should you lack cardiovascular fitness, your golf performance is likely to suffer and you will fatigue easily. Fatigue from any source negatively affects fine motor coordination. If you get tired walking from hole to hole, you are not likely to play at your best.

It is best to do cardiovascular exercise after your strengthening exercises should you want to do both in the same session. If you choose programming Option C, it is best to do your cardiovascular exercise on the days between your strength training sessions. It is generally recognized that three 30-minute sessions per week are sufficient to maintain cardiovascular health.

It is recommended that you see your doctor or an exercise physiologist to have your cardiovascular system tested. It is incredibly useful to know where you stand regarding your health. If you need more cardiovascular exercise, your medical professional will be able to tell you.

If you are able to consult a CHEK Golf Biomechanic or C.H.E.K Practitioner to get your program individually customized, they can design an exercise program with cardiovascular conditioning that is specific to your personal needs. Cardiovascular exercise must be used carefully if you take your golf seriously because most cardiovascular exercise is cyclical in nature. Cyclical exercise sedates the nervous system, which is opposite of what the player needs for optimal performance! The C.H.E.K Institute-trained Professional can show you how to perform "functional cardio" exercises that are both good for your game and your health.

CHAPTER FIVE

PROGRAMMING OPTION A - DAILY PLANNING WORKSHEET

TRAINING PHASE: Phase I – Neuromuscular Isolation Exercises
OBJECTIVE: Increase brain / muscle communication
DATES: 6 weeks

Neuromuscular Isolation – Phase I

Exercise	Rest	Intensity	Reps	Tempo	Sets
Stretch & Warm Up Set First					
Monday					
Prone Cobra	<1:00		1-8	30/15 secs	1-2
Tuesday					
Horse Stance Vertical/Horizontal	1:00		10 each	10/10	1-3
Wednesday					
Swiss Ball Neck Training*	1:00	<40% max	1-3 each	Hold 30 secs	1-3
Thursday					
Four Point TVA Trainer	<1:00		10	10/10	1-3
Friday					
Greg Johnson's Window Washer	L/R		12-20 each	Slow	1-3 each
Saturday					
Swiss Ball Hyper-Extension	1:00		8-12	333	1-3
Sunday					
Lower Abdominal #1-3	<1:00	+30 mmHg	**	S/M/F	1-3 each

* Perform one rep of each of the SB Neck Training exercises in turn. Hold each position at <40% intensity until neck fatigues and then change positions. As conditioning improves, repeat circuit 1-3 times.

** Lower Abdominal #1 is performed for ten repetitions per set.
   Lower Abdominal #2 is performed for 12-20 repetitions per set.
   Lower Abdominal #3 is performed for 8-12 repetitions per set.
   Select the appropriate exercise for your level of conditioning.

PROGRAMMING OPTION B – DAILY PLANNING WORKSHEET

TRAINING PHASE: Phase I – Neuromuscular Isolation Exercises
OBJECTIVE: Increase brain / muscle communication
DATES: 4-6 weeks

Neuromuscular Isolation – Phase I

Exercise	Rest	Intensity	Reps	Tempo	Sets
Stretch & Warm Up Set First					
Monday					
Horse Stance Vertical/Horizontal	1:00		10 each	10/10	1-3
Prone Cobra	<1:00		1-8	30/15 secs	1-2
Wednesday					
Swiss Ball Neck Training*	1:00	<40% max	1-3 each	Hold 30 secs	1-3
Four Point TVA Trainer	<1:00		10	10/10	1-3
Friday					
Greg Johnson's Window Washer	L/R		12-20 each	Slow	1-3 each
Swiss Ball Hyper-Extension	1:00		8-12	333	1-3
Sunday					
Lower Abdominal #1-3	<1:00	+30 mmHg	**	S/M/F	1-3 each

* Perform one rep of each of the SB Neck Training exercises in turn. Hold each position at <40% intensity until neck fatigues and then change positions. As conditioning improves, repeat circuit 1-3 times.

** Lower Abdominal #1 is performed for ten repetitions per set.
   Lower Abdominal #2 is performed for 12-20 repetitions per set.
   Lower Abdominal #3 is performed for 8-12 repetitions per set.
   Select the appropriate exercise for your level of conditioning.

PROGRAMMING OPTION C - DAILY PLANNING WORKSHEET

TRAINING PHASE: Phase I – Neuromuscular Isolation Exercises
OBJECTIVE: Increase brain / muscle communication
DATES: 4 weeks

Neuromuscular Isolation – Phase I

Exercise	Rest	Intensity	Reps	Tempo	Sets
Stretch & Warm Up Set First					
Training Day 1&3					
Horse Stance Vertical/Horizontal	1:00		10 each	10/10	1-3
Swiss Ball Neck Training*	1:00	<40% max	1-3 each	Hold 30 secs	1-3
Four Point TVA Trainer	<1:00		10	10/10	1-3
Prone Cobra	<1:00		1-8	30/15 secs	1-2
Training Day 2 & 4					
Horse Stance Vertical/Horizontal	1:00		10 each	10/10	1-3
Greg Johnson's Window Washer	L/R		12-20 each	Slow	1-3 each
Swiss Ball Hyper-Extension	1:00		8-12	333	1-3
Lower Abdominal #1-3	<1:00	+30 mmHg	**	S/M/F	1-3 each

* Perform one rep of each of the SB Neck Training exercises in turn. Hold each position at <40% intensity until neck fatigues and then change positions. As conditioning improves, repeat circuit 1-3 times.

** Lower Abdominal #1 is performed for ten repetitions per set.
   Lower Abdominal #2 is performed for 12-20 repetitions per set.
   Lower Abdominal #3 is performed for 8-12 repetitions per set.
   Select the appropriate exercise for your level of conditioning.

FUNCTIONAL EXERCISE - PHASE II
Neuromuscular Integration Exercises

Neuromuscular Integration Exercises are similar to Neuromuscular Isolation Exercises but they are more complex. With Neuromuscular Integration Exercises, you will have to stabilize, and/or move more segments of your body at one time, the only exception being the Grip Trainer exercises. These exercises serve as progressions from Neuromuscular Isolation Exercises and as prerequisites for Dynamic Stability Exercises.

The Neuromuscular Integration Exercises can be performed for four to six weeks. As with the Neuromuscular Isolation programming options, should you choose Option A or B, it is recommended that you stay with the exercises for six weeks to allow enough exposure for the learning process to take place.

Before beginning Phase II, study the exercises and their descriptions very carefully. You must perform them with accurate form in order to get results that transfer into lower scores per game!

Phase II Equipment Needs

To effectively perform the exercises in Phase II you will need the following equipment:

- One Swiss ball*

- One wooden dowel rod six feet long by $1^{1}/_{4}$ to $1^{3}/_{8}$" diameter which can be purchased at any building supply store (closet rod). A PVC pipe of the same dimensions will also work.

- One pair of dumbbells ranging from 5-50 lbs., depending on how strong you are. You can also use a pair of 1-gallon plastic milk jugs filled with water.

- One Power Web*

* See Resources section on page 219

CHAPTER FIVE

PHASE II EXERCISE DESCRIPTIONS

Exercise	Rest	Intensity	Reps	Tempo	Sets
Supine Hip Extension: Feet on Ball	1:00		8-12	333	1-3

The Supine Hip Extension: Feet on Ball is an excellent exercise providing several benefits:

- Conditions the core musculature as stabilizers
- Trains the hamstrings, butt and back muscles to work together
- Improves balance
- Improves coordination
- Improves postural endurance

- Lie on your back and place your feet on the ball (Figure 5.21A). Start with the feet wide and narrow them as you progress.

- Put your arms out to your side with your *palms facing up*. If this is your first time using a Swiss Ball and your balance is not refined yet, you may want to place your calves on the ball and your arms straight out to your sides at a 90° angle to your trunk.

- From the start position, extend the hips into the air over the count of three until your ankle, hip and shoulder all line up (Figure 5.21B). Hold for three seconds and then lower for three seconds.

- As you become more proficient at the exercise, you can move your hands closer to your body. Also, the exercise can be made more challenging by placing less of your legs on the ball as well as by moving your arms closer to your body, eventually placing them across your chest. When you really feel frisky, you can attempt the exercise with your arms across your chest and just your heels on the ball!

Figure 5.21
Supine Hip Extension: Feet on Ball

FUNCTIONAL EXERCISE

Exercise	Rest	Intensity	Reps	Tempo	Sets
Forward Ball Roll	1:00		8-12	333	1-3

The Forward Ball Roll strengthens the abdominals, hip flexors and shoulder extensors. This exercise is very good for golfers as it improves the golfer's ability to stabilize his or her spine and teaches some of the major muscle groups to work together. This exercise is useful for the golfer wanting to add distance to his or her drive, as the muscles conditioned here are very important to the drive action of swinging a golf club.

- Find a comfortable surface, such as an exercise mat, carpeted floor or grass. From a kneeling position, place your forearms on the ball with your palms facing each other (Figure 5.22A).

- Place a dowel rod on your back. The goal is to maintain good spinal alignment as you roll forward. Good spinal alignment is indicated by not exaggerating any of your spinal curvatures, which often causes the rod to fall!

- Take a deep breath and draw the navel toward your spine just enough to slim your waistline slightly.

- Begin rolling forward, moving from the hip and shoulder joints equally (Figure 5.22B). The movement should terminate the instant you feel you are going to lose spinal alignment, but don't allow yourself to lose alignment. You will know you are losing spinal alignment if the curves in your spine increase and the rod starts to fall off!

- The key to good performance with the Forward Ball Roll is to stop just short of the point at which you start to lose your ability to hold the spine in perfect alignment. Pushing past that point will only serve to magnify muscle imbalances and decrease golf performance.

- Having found your endpoint, you need to adjust your tempo so that it takes three seconds to reach the end point, hold for three seconds, and then roll back for three seconds, exhaling as you return. This allows adequate strengthening of both the large prime mover muscles and the abdominal stabilizer muscles.

Figure 5.22
Forward Ball Roll

CHAPTER FIVE

Exercise	Rest	Intensity	Reps	Tempo	Sets
Horse Stance Alphabet	1:00		Max	Slow	1-3

- From the same start position described for the Horse Stance Horizontal (Figure 5.6), place the dowel rod along the spine as seen in Figure 5.23.

- With the arm 45° to the side and the thumb up, use the extended leg to draw letters of the alphabet. Start with small letters of 4-6 inches high and progress to larger letters as you are able to stabilize your core and keep the dowel rod in place.

Figure 5.23
Horse Stance Alphabet

- **When performing the exercise, it is important to make sure the following checkpoints are met:**

  - The head and neck should stay in line with the spine. The head should not drop down nor look up at any time.

  - Elbow of support arm should point directly backward, not to the side.

  - The arm that is up should maintain an angle of 45° off the midline of the body at all times.

  - The shoulders and hips should remain parallel with the floor at all times.

  - There should be no significant movement of the low back. The movement of the leg needed to draw the letters of the alphabet should come from the hip.

  - The lower leg should move as a unit with the thigh. It is not good technique to just use the lower part of the leg.

  - Draw the navel toward the spine throughout.

- Perform as many repetitions as possible with perfect form before switching sides. This is indicated in your Reps column as Max. When you can perform the entire alphabet on either side with perfect form, add a 1 lb. weight to each wrist and a 3 lb. weight to each ankle.

FUNCTIONAL EXERCISE

Exercise	Rest	Intensity	Reps	Tempo	Sets
Prone Bridge	1:00		Max	S-M	1-3

The Prone Bridge (Figure 5.24) is challenging, so don't be disappointed if you fall off the ball a few times. To be safe, make sure you do this exercise on a mat, carpet or grass. If you cannot do the exercise initially, try performing it with your feet on a bench instead of a ball. As your strength and stability develop, try using a Swiss ball again.

- Grasp the ball between your shins and assume a push-up position. Hold your body in perfect alignment with ankles, hips, shoulders and head all in the same horizontal plane (Figure 5.24A).

- Once you feel balanced, take one hand off the ground for just a second. Practice just holding your alignment for short intervals with only one hand on the ground.

- As you improve, begin raising the arm forward.

- When you can get the arm out in front of you, begin to move it from the front position in an arc from over your head to your side and back to the floor again (Figure 5.24B).

- Alternate between your left and right arms as many times as you can (Max in the Reps column) without losing good postural alignment.

Figure 5.24
Prone Bridge

CHAPTER FIVE

Exercise	Rest	Intensity	Reps	Tempo	Sets
Frontal Plane Static Lean	L/R		1-2 each	Max	1-2

The Frontal Plane Static Lean exercise is designed to improve the strength and endurance of the muscles that resist side bending. You may have experienced the one-sided back fatigue that comes when you carry your clubs over one shoulder. This fatigue is primarily from the oblique abdominal muscles and the deep back muscle called the quadratus lumborum.

These important muscles have a working relationship with the muscles on the side of your hip and the groin muscles. These muscles all work together so you can walk without dragging your swing leg and they also greatly assist in generating the power to swing your golf clubs. The Frontal Plane Static Lean exercise will improve all these functions as well as improve your ability to carry a briefcase, suitcase, or groceries without hurting your back.

Figure 5.25
Frontal Plane Static Lean

- Find a secure place to anchor your feet as seen in Figure 5.25. The wall works well if you can't find any secure objects or a partner.

- For the beginner, the Swiss Ball should be placed with the apex at, or slightly above belt-line level.

- Try the exercise with your arms at your side (Figure 5.25A).

- As you become stronger and are able to hold the position for three minutes or more, progress the exercise by placing your hands across your chest, by your ears (Figure 5.25B), then over your head (Figure 5.25C). Progress only when you are sure you can handle the added stress. Your arms and shoulders weigh about 14% of your body weight and raising them up to your chest or head makes a big difference on how hard the muscles must work to hold the exercise position. Even if you think you are strong, start at the beginner level with the ball above the belt-line and the arms at your side!

- Hold the body in good alignment at about a 45° angle to the floor for as long as you comfortably can.

- Start with only one repetition per side on your first session or two.

FUNCTIONAL EXERCISE

Exercise	Rest	Intensity	Reps	Tempo	Sets
Bent Over Row	1:00	-2 reps	8-12	203	1-3

The Bent Over Row is an excellent exercise for improving the postural endurance of your back, butt and hamstring muscles as well as strengthening your shoulder girdle.

- To find your optimal Bent Over Row position, assume the ready position of a shortstop in baseball – knees bent slightly with arms straight and palms resting on the knees. This will put your trunk in the correct position, which is at an angle of 45-60° as shown in Figure 5.26. This position is slightly lower than your address position.

- Hold dumbbells of light to moderate weight in your hands. Imagine that you are a puppet and your elbows are being lifted toward the sky by strings. Be very careful not to lift with your hands; only use enough grip to hold the weights. Think of lifting the weights from your shoulders to recruit the muscles properly.

- Draw the weights up to your side, keeping the elbows straight out rather than pulling them back toward the hips. If you allow the elbows to go back toward the hips, you will not be strengthening the shoulder girdle muscles that help maintain good postural alignment and stabilize the shoulder during the golf swing. Avoid shrugging your shoulders.

- If you have chosen the correct weight, then you should be able to complete twelve good repetitions and feel like you could do two more good ones if you had to. This will allow a buffer zone so that as you tire you don't drop below eight reps and you always stop before your form fails.

Figure 5.26
Bent Over Row

Start

Finish

CHAPTER FIVE

Exercise	Rest	Intensity	Reps	Tempo	Sets
Grip Trainer	L/R	#	12-20	Slow	2-6

Choose appropriate resistance of Power Web.

Golfers commonly have problems with the forearm muscles used to hold the club, therefore we must condition these grip muscles. The Power Web is an excellent device for developing grip strength. They come in different levels of resistance (indicated by # in the intensity column), so once you can do six sets of finger flexion and extension exercises with the low level Power Web, you can progress to the next level.

- Place your fingers in the holes just outside your grip width.

- With your arm held straight out to the side, begin squeezing and opening the grip at a slow tempo (Figure 5.27A).

- If your grip fatigues between 12-20 reps, you have the right resistance. If not, you can change the resistance by opening the grip more to increase the difficulty, closing the grip a little before sticking your fingers through the web to decrease the resistance, or by using a different density Power Web.

- Immediately after working the muscles that close your hand (finger flexors), stick your fingers through the web close together so you have to work to open them (finger extensors) (Figure 5.27B).

- Find the right level of resistance to allow opening the fingers against the Power Web 12-20 times.

- Once you have exercised the finger flexors and extensors of one arm, switch to the other arm. This process should be repeated 1-6 times per side. Start with only one set, adding sets slowly to prevent muscle strain.

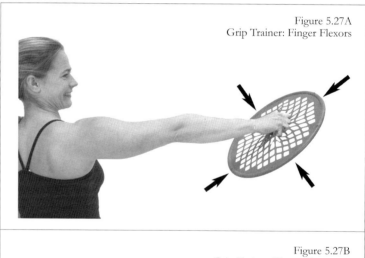

Figure 5.27A
Grip Trainer: Finger Flexors

Figure 5.27B
Grip Trainer: Finger Extensors

PROGRAMMING OPTION A DAILY PLANNING WORKSHEET

TRAINING PHASE: Phase II – Neuromuscular Integration Exercises
OBJECTIVE: Increase brain / muscle communication
DATES: 4-6 weeks

Neuromuscular Integration – Phase II

Exercise	Rest	Intensity	Reps	Tempo	Sets
Stretch & Warm Up Set First					
Monday					
Supine Hip Extension: Feet on Ball	1:00		8-12	333	1-3
Tuesday					
Forward Ball Roll	1:00		8-12	333	1-3
Wednesday					
Horse Stance Alphabet	1:00		Max*	Slow	1-3
Thursday					
Prone Bridge	1:00		Max**	S-M	1-3
Friday					
Frontal Plane Static Lean	L/R		1-2 each	Max***	1-2
Saturday					
Bent Over Row	1:00	-2 reps	8-12	203	1-3
Sunday					
Grip Trainer	L/R	#	12-20	Slow	2-6

Max*: Perform as many repetitions as possible on one side with perfect form, then switch sides.
Max**: Alternate arms as many times as possible. Do not go over 2 minutes.
Max***: Hold position for as long as possible, up to 3 minutes per side.
#: Choose appropriate resistance of Power Web.

CHAPTER FIVE

PROGRAMMING OPTION B DAILY PLANNING WORKSHEET

TRAINING PHASE: Phase II – Neuromuscular Integration Exercises
OBJECTIVE: Increase brain / muscle communication
DATES: 4-6 weeks

Neuromuscular Integration – Phase II

Exercise	Rest	Intensity	Reps	Tempo	Sets
Stretch & Warm Up Set First					
Monday					
Supine Hip Extension: Feet on Ball	⇩1:00		8-12	333	1-3
Forward Ball Roll	⇩⇨⇧		8-12	333	1-3
Wednesday					
Bent Over Row	⇩1:00	-2 reps	8-12	203	1-3
Horse Stance Alphabet	⇩⇨⇧		Max*	Slow	1-3
Friday					
Prone Bridge	⇩1:00		Max**	S-M	1-3
Frontal Plane Static Lean on SB	⇩⇨⇧		1-2 each	Max***	1-2
Sunday					
Grip Trainer	L/R	#	12-20	Slow	2-6

Max*: Perform as many repetitions as possible on one side with perfect form, then switch sides.
Max**: Alternate arms as many times as possible. Do not go over 2 minutes.
Max***: Hold position for as long as possible, up to 3 minutes per side.
#: Choose appropriate resistance of Power Web

PROGRAMMING OPTION C DAILY PLANNING WORKSHEET

TRAINING PHASE: Phase II – Neuromuscular Integration Exercises
OBJECTIVE: Increase brain / muscle communication
DATES: 4 weeks

Neuromuscular Integration – Phase II

Exercise	Rest	Intensity	Reps	Tempo	Sets
Stretch & Warm Up Set First					
Training Days 1 & 3					
Bent Over Row	⇩1:00	-2 reps	8-12	203	1-3
Horse Stance Alphabet	⇩		Max*	Slow	1-3
Forward Ball Roll	⇩⇨⇧		8-12	203	1-3
Grip Trainer	L/R	#	12-20	Slow	2-6
Training Days 2 & 4					
Prone Bridge	⇩1:00		Max**	S-M	1-3
Supine Hip Extension	⇩		8-12	333	1-3
Frontal Plane Static Lean	⇩⇨⇧		1-2 each	Max***	1-2

Max*: Perform as many repetitions as possible on one side with perfect form, then switch sides.
Max**: Alternate arms as many times as possible. Do not go over 2 minutes.
Max***: Hold position for as long as possible, up to 3 minutes per side.
#: Choose appropriate resistance of Power Web

CHAPTER FIVE

FUNCTIONAL EXERCISE – PHASE III
Dynamic Stability Exercises

Now that you have traded your canoe in for a catamaran, you can begin conditioning Dynamic Stability. Dynamic Stability refers to the ability to maintain optimal joint alignment at all times during each and every golf swing.

To highlight this important concept, consider the analogy of the spinning top. The golfer's body rotates as he/she swings a club, just as a top must rotate to stay aligned along its axis of rotation. The top spins perfectly, balanced around its center of gravity, falling only when inertial energy declines to critical levels (Figure 5.28). Should you bend the stem of a top and spin it, the top will not maintain a central axis of rotation. It will flail about, throwing inertial energy in all directions. The same is true of the golfer who cannot maintain the "instantaneous axis of rotation" or alignment of the spine and extremity joints. They will be unable to reproduce a consistent, accurate rotation or golf swing.

To improve dynamic stability during your golf swing it is essential that you learn to maintain optimal alignment of the head, spine and extremity joints while performing the swing. The only way this goal can be achieved is through a combination of optimal flexibility, static stability and dynamic stability. Dynamic stability in the golfer is easily identified as the ability to integrate movements of the hip-pelvis region with the trunk and shoulder girdle over a progressively decreasing base of support, just as during the golf swing.

Figure 5.28
Good Postural Alignment Produces Optimal Swing

The exercises for improving dynamic stability are aimed at integrating these muscle joints. It is very important for the golfer to achieve optimal flexibility, static and dynamic stability prior to attempting the golf strengthening and golf power exercises in Chapters Six and Seven.

FUNCTIONAL EXERCISE

Due to the large variability of fitness levels and ages among active golfers, some of the exercises presented in this phase of your conditioning program may seem very advanced. The exercise descriptions explain how to increase or decrease the complexity of each exercise. Exercises with a high level of difficulty are indicated with this symbol * in the Daily Planning Worksheets.

At this point in the book, you have probably come to realize that the Whole in One approach to golf conditioning is much more comprehensive than the traditional gym program or programs presented in other books.

Some exercises will require the purchase of special equipment. Should you not wish to purchase the equipment, simply exclude that exercise from the program. Before you make the decision to drop an exercise that will most likely improve your game, look down at your shoes. Why do you wear those shoes? Now, go look at your clubs. Why did you buy good clubs, not just average clubs, and good shoes, not average shoes? It's because you want the best. It's simple then; keep wanting the best and do what it takes to prepare properly so you can get the best from those good golf shoes and clubs!

Once you have completed four to six weeks of Dynamic Stability Exercises, you will have significantly improved joint function, reduced postural sway and enhanced nervous system function. Below the exercise explanations you will find three programming options. Should you choose Option A or B, it is again best to stick to the program for six weeks to allow your body time to learn from the exercises.

As you complete your Dynamic Stability program you will no doubt continue to experience improved golf performance and decreased likelihood of golf-related back pain. Work diligently and you will be prepared for your Golf Strength Program!

THE YOUNG OR FIT GOLFER READY FOR A CHALLENGE

It is very important for all golfers to follow the program as outlined up to this point. Many top football, baseball, basketball, rugby, surfing, tennis and hockey players and other elite athletes have completed very similar phases of progression. The key areas addressed up to this point are common weaknesses among most athletes and non-athletes alike.

If you have done well with the exercises and feel up to a challenge, you can actually perform Dynamic Stability: Option A and Strength Training: Option A at the same time. The way to do this is to perform one Dynamic Stability exercise at the end of every strength training session. This will allow concurrent development of strength and dynamic stability. Should you choose this option, it is very important to rest for five minutes between finishing your strength exercises and starting your dynamic stability exercises.

CHAPTER FIVE

Phase III Equipment Needs

- Burst resistant Swiss ball*

- Dumbbells ranging from 5-30 lbs. depending on your strength level

- One 6' x $1^3/_8$" wooden dowel rod, or PVC piping as in Phases I & II

- One Power Web*

OPTIONAL EQUIPMENT:

- Fitter Wobble Board*

- The Pro Fitter*, an unique device that develops balance and coordination

* See Resources section on page 219.

Sizing Your Swiss Ball

Swiss balls should be firm when used for exercising. Sit on top of the inflated ball. Your thighs should be parallel to the floor. If you suffer from backache, the ball can be a little bigger, so your thighs are just above parallel.

- If your height is under 5'2" you will need a 45cm / 18" ball.

- If your height is 5'2" to 5'8" you will need a 55cm / 21½" ball.

- If your height is 5'9" to 6'2" you will need a 65cm / 25" ball.

- If your height is 6'4" to 6'9" you will need a 75cm / 29½" ball.

FUNCTIONAL EXERCISE

PHASE III EXERCISE EXPLANATIONS

Exercise	Rest	Intensity	Reps	Tempo	Sets
Supine Hip Extension: Knee Flexion	1:00		8-12	Slow	1-3

The Supine Hip Extension: Knee Flexion is an advancement on the Supine Hip Extension: Feet On Ball exercise performed in Phase II. This exercise is very useful for improving your hamstring strength and coordination at both the knee and hip joints.

This exercise not only conditions the hamstrings, but it also trains the low back and abdominal musculature. As you will see, when your feet are on the ball and you extend the hip while flexing the knee, the ball will try to roll to one side if your foot pressure is uneven. To keep from slipping off the ball, you must use your abdominal musculature to stabilize, just as you must use you abdominal musculature to stabilize the trunk and pelvis to maintain your swing axis during the golf swing.

- Lie on your back with your arms outstretched and palms up. Place the lower calf region of both legs on the ball.

- Start the exercise just as you did during the Supine Hip Extension: Feet on Ball excrise, slowly raising your hips up until they are in line with your shoulders and ankles (Figure 5.29A).

- Now flex your knees (Figure 5.29B). It is very important that you don't let your hips drop when flexing the knees. If you cannot flex your knees without the hips dropping, it indicates that you lack hamstring strength and should continue with the Supine Hip Extension: Feet on Ball version performed in Phase II.

- To make the exercise more challenging, move the ball farther away from your body.

Figure 5.29
Supine Hip Extension: Knee Flexion

For those of you who can flex your knees without your hips dropping, slowly reverse the process by slowly straightening your knees and then lowering the pelvis to the start position. The exercise is very challenging, but very rewarding!

Progressing the Exercise

If you master the exercise before the end of Phase III and want an additional challenge, progressively bring your arms closer to your body. This will reduce your base of support, and increase the demand on your spinal and abdominal musculature in order to maintain your balance.

Precautions

It is very important that you use a ball that is the correct size for you. If you are using a friend's ball, or any ball that is too big, you could place unnecessary stress on your neck doing this exercise.

If you have any history of neck discomfort, get permission to do this exercise from your doctor before attempting it. A modification that can be very useful for someone with neck discomfort is to use either a 45 cm Swiss ball or a basketball, both of which significantly reduce the stress placed on the neck during the exercise. If you still experience neck discomfort while using a basketball, consult your physician or an orthopedic physical therapist.

Exercise	Rest	Intensity	Reps	Tempo	Sets
Prone Jack Knife	1:00		8-12	202	1-3

The Prone Jack Knife exercise is used to strengthen the golfer's hip flexors, abdominal muscles and shoulder girdle. This exercise helps train the golfer to disassociate hip movements from back movements. Allowing movement that should be isolated to the hip to overflow into the back is a common source of back pain for many golfers.

If the golfer is weak in the hip flexors and cannot differentiate hip movements from back movements, then there is a significantly increased chance of over-using the low back when swinging a golf club. Performing the Prone Jack Knife with perfect form will aid in re-establishing normal movement patterns and strengthen areas in which golfers are commonly weak.

- In a push-up position, place your feet on the ball (Figure 5.30A).
- Hold your spine straight and maintain head and neck alignment.
- Draw your legs under your body over the duration of two seconds, without changing the curves in the back (Figure 5.30B).
- Return to the start position over the duration of two seconds.
- Repeat the exercise for the prescribed number of repetitions.

If you find the exercise hard to perform, you can modify it in the following ways:

- Place more of your leg on the ball, for example half or all of your shins. This will reduce the lever arm acting against your abdominal muscles as well as reduce the load supported by your arms.

- Another Swiss ball can be used to support the torso. This is done by deflating a Swiss ball to the point that it gives just enough support to allow completion of the exercise. As you become stronger, deflate the support ball a little each training session. This will help to progressively increase the demand on your musculature, making you stronger.

Figure 5.30
Prone Jack Knife

CHAPTER FIVE

Exercise	Rest	Intensity	Reps	Tempo	Sets
Supine Lateral Ball Roll	1:00		6-8	1 sec hold	1-3

The Supine Lateral Ball Roll is a "BIG BANG" exercise. It improves many things in your body at one time. For example, performing this exercise with correct technique can significantly improve the following:

- Posture
- Balance
- Coordination
- Shoulder strength girdle
- Core strength
- Righting reflexes
- Agility
- Endurance
- Golf performance!

- Lie on your back on a Swiss Ball (Figure 5.31A). Position your body so that your head is comfortably supported on the ball, as well as the area between your shoulder blades. Extend your hips upward until your knees, hips and shoulders are all in the same horizontal plane.

- Place your tongue on the roof of your mouth just behind your front teeth, a position that can be found by swallowing (this is called the physiological rest position of the tongue).

- Extend your arms out and turn the hands so the palms are facing upwards. Place a dowel rod in your hands.

- Begin to roll laterally. Throughout the exercise, hold the alignment of your body exactly as it was before moving, with the exception that you may move your feet in a small shuffle to keep them in alignment with the body as you shift laterally (Figure 5.31B).

- Go only to the point that you can hold the alignment for the count of "one thousand and one," then return to the opposite side, repeating the sequence. To perform the prescribed 6-8 repetitions per side, you must only move as far laterally as you can and still last long enough to perform the full set.

Figure 5.31
Supine Lateral Ball Roll

Note: It is very important to warm-up for the exercise by doing a set of easy reps.

As you get stronger, the Supine Lateral Ball Roll can be made more challenging by simply going farther out off the ball. You can also make the exercise tougher by holding the end position for a longer period of time on each side. Be very careful not to overdo it, as this exercise works many muscles quite hard and can leave the over-aggressive exerciser sore for a couple days!

CHAPTER FIVE

Exercise	Rest	Intensity	Reps	Tempo	Sets
Prone Twister: Feet on Ball	1:00		6-10	S-M	1-3

The Prone Twister: Feet on Ball will help golfers develop explosive coil strength, adding distance to their drives!

- Place your feet on the ball from the push-up position. Position your feet so that you can pinch the ball between them in order to control the ball.

- Holding your body so that your legs, trunk and head are in the same horizontal plane, begin twisting the lower body to the left and the right slowly and gently (Figure 5.32). To warm-up, you should go very easy, never exerting greater than about 60% of your training intensity.

- Once you are warm, perform the exercise at a slow tempo.

- When you can perform the prescribed number of sets with perfect form, holding your body well-aligned in the horizontal plane, you can speed the tempo up to moderate (202).

Figure 5.32
Prone Twister: Feet on Ball

FUNCTIONAL EXERCISE

Exercise	Rest	Intensity	Reps	Tempo	Sets
Kneeling on Swiss Ball	1:00		1-3	Hold 30 secs	4-8

Kneeling on the Swiss Ball is a fun exercise with numerous benefits for the golfer, including improvement of:

- Balance
- Righting Reflexes
- Tilting Reflexes
- Coordination
- Dynamic Posture

- Place your hands and knees on a Swiss ball (Figure 5.33A) and slowly rock forward until you are balancing on the ball with your hands and knees (Figure 5.33B).

- From there, let go with your hands and raise up onto your knees (Figure 5.33C). It will help to use as much of your shins as possible to balance, as your shins create a larger base of support than just your knees. You may also hold onto the ball with your feet for added stability.

- Initially, you should just practice kneeling for as long as possible without coming off the ball. When you get to the point where you can kneel on the ball for thirty seconds straight, several sets in a row, you can then progress to the next phase of the exercise.

Figure 5.33A
Kneeling on Swiss Ball

A B C

CHAPTER FIVE

Kneeling Swiss Ball Golf Swing

Now that you have developed your balance to the point of being comfortable on your knees and not falling off the ball, you are ready for the next challenge.

- Kneel on a Swiss ball and assume an address position, pretending to hold a golf club.

- Perform half a backswing, then the downswing and finally follow-through to 50% range of motion. Practice this exercise for as long as you can without feeling like you are losing good form or for up to thirty seconds.

- Once you can do several sets with good form, you can progress the exercise by actually holding a small dumbbell or a golf club in your hands. The added weight in your hands will constantly alter the location of your center of gravity, making the exercise much more challenging.

Note: The Swiss ball should be firmly inflated. If the ball is too soft, the exercise will not be nearly as beneficial.

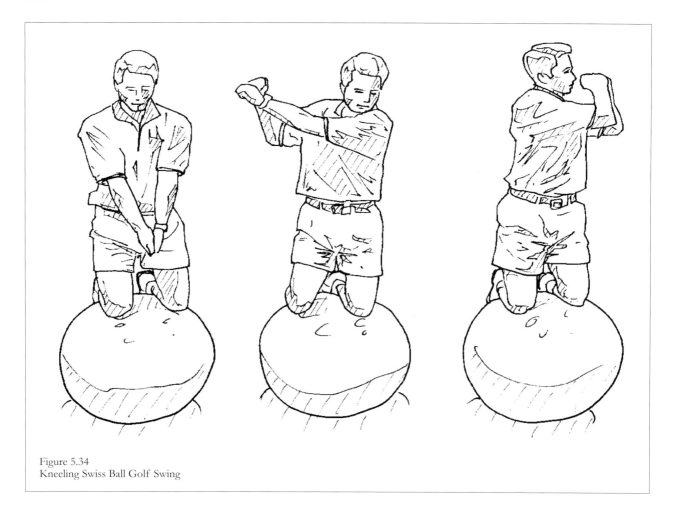

Figure 5.34
Kneeling Swiss Ball Golf Swing

Exercise	Rest	Intensity	Reps	Tempo	Sets
Supine Russian Twist	1:00	-2 reps	6-10 each	S-M	1-3

The Supine Russian Twist is an excellent exercise for developing core strength. It also improves the strength of your pelvic girdle, which is very important to your golf swing. Another important feature about the exercise is that the feet are on the ground while the hip and trunk musculature are working together to perform trunk rotation. Having the feet on the ground improves the brain's ability to transfer the strength gained from this exercise to the golf swing. The foot and ankle provide the brain with very important information about body direction, weight shift, associated movement and other vital pieces of information.

- Place your back and head on the ball so that your head is comfortably supported and your shoulders are across the apex of the ball. Lift the hips so that they are in line with the knees and shoulders in the horizontal plane (Figure 5.35A). Reach up and clasp the hands together.

- Place your tongue on the roof of the mouth to increase the activation of your neck flexors, which stabilize your head and neck.

- Begin rotating the trunk as far as possible to one side and then the other using a slow tempo (Figure 5.35B+C). Move slowly for several repetitions, progressively increasing the range of motion. After about six repetitions from side-to-side, you can speed up the tempo to moderate. As you rotate your arms and trunk side-to-side, be sure to keep your hips up. Make sure your hips do not drop. Emphasis should be placed on trunk rotation with this exercise. The first set should not cause you to exert yourself, it is just a warm-up.

- You can continue with no added weight and go through your prescribed sets, or you can hold a small medicine ball 1-6 lbs. (0.5-3 kg.). This exercise may cause some post-exercise soreness, so go easy at first. Only add extra resistance if you do not get sore after your first work-out.

Figure 5.35
Supine Russian Twist

CHAPTER FIVE

Exercise	Rest	Intensity	Reps	Tempo	Sets
Fitter Wobble Board	1:00	##	1-3	Hold 30 secs	1-8

##: Board level 1, 2 or 3

This exercise is intended to be performed on a wobble board, which uses a dome or half sphere as the pivot axis. A rocker board, which is unstable in only one plane of motion, may also be used. These are more suitable to senior golfers and those who are not confident about their balance abilities.

The Fitter Wobble Board exercise is another very effective method for developing balance (see Figure 5.36), particularly for golfers who have suffered from back pain in the past, or those who have had back surgery. After back surgery, or in the presence of back pain, the muscles frequently develop faulty recruitment patterns, which only perpetuate the dysfunction and chronic low back pain. Balancing on a wobble board or rocker board causes momentary loss of balance, which is perceived as a threat by the brain. In response to the potential threat, the brain overrides faulty recruitment patterns. This is when the low back muscles that have not been working correctly get a jump-start. With repetitive exposure, the back muscles begin working again in many cases.

Figure 5.36
Fitter Wobble Board

When using a Fitter Wobble Board, you can adjust the height of the pivot, making the exercise harder as the pivot gets higher.

- To begin, set the pivot to the lower level. The goal of the exercise is to stand on the board for as long as possible without letting the edge of the board touch the ground.

- It is very important to do the exercise in front of a mirror so you can keep an erect posture while balancing on the board. If you look at the feet too much, it can encourage poor postural development.

FUNCTIONAL EXERCISE

If the exercise seems difficult in the beginning, don't be discouraged. It is challenging! To make it a little easier for beginners, try putting the wobble board on an exercise mat. This slows the board down, making it easier to find a balance point. If you don't have an exercise mat available, carpet will do the trick.

To get the most of your balance board training, make only as many attempts as you can in thirty seconds to keep the edges of the board off the floor. If you seem to be getting worse as time goes by, stop. As indicated in the Reps column, you should perform 1-3 attempts for no more than thirty seconds each. Doing more than this per set usually results in loss of performance, not better performance. After the prescribed rest period of one minute, perform another set. Again, do only as many sets as you can do with good quality; if you seem to keep falling off the board, stop for the day. Your nervous system is probably tired!

Exercise	Rest	Intensity	Reps	Tempo	Sets
Pro Fitter	1:00		Max	S-M-F	1-8

The Pro Fitter is a balance and proprioceptive device that was originally developed for skiers. Over the years, the Pro Fitter has become very popular with rehabilitation specialists and conditioning coaches for many sports. For the golfer, the Pro Fitter is an excellent method of improving balance, coordination, reflexes, and strength. Depending on which way the Pro Fitter is used (Figures 5.37A-B), you will develop strength, balance and coordination in different planes of movement.

Figure 5.37A
Pro Fitter - Lateral

- Start by developing skill with lateral movements (Figure 5.37A).

- Stand on the foot supports with good upright posture, preferably in front of a mirror. Beginners may want to use one or two support poles.

- The exercise is simply to work at moving your lower body laterally while keeping your upper body as still as possible. Using the ski pole supports is very helpful.

- As you become more proficient at the exercise, reduce the support from two ski poles to one, and finally to none.

CHAPTER FIVE

Figure 5.37B
Fitter - Sagittal Plane

After a brief period of experimentation, you will determine which resistance level to use on your Pro Fitter. The resistance is easily adjusted by either adding or subtracting stretch cords that attach underneath the trolley unit.

Try just playing with the Pro Fitter! This type of training is, and should be, fun! To get the most from your nervous system, play for periods of no more than one minute without taking a rest. As your skill level increases and your nervous system adapts, you can progressively add sets. When you have mastered lateral movement with no poles for support, you are ready to try front-to-back movement, a sagittal plane movement (Figure 5.37B). (If you want to learn more challenging Pro Fitter exercises, order the High Performance Core Conditioning DVDs*).

Front to back (sagittal) movement is much more challenging than lateral movement. Again, you will want to start the learning process with a ski pole or two, depending on how aggressive you are.

- Stand on the Pro Fitter in much the same way as you would stand on a skateboard or water ski. You can attach the seat pad to the foot rails to provide a firmer surface to stand on.

- Rock gently back and forth. As before, the goal is to keep the lower body moving independently from the upper body. You may find that the front-to-back movements are easier to perform with two or three cords attached.

- Having advanced to the level of being able to perform sagittal plane movements on the Pro Fitter, you should now begin to mix the two movements in your training sessions. The best way to do this is randomly, as it has been shown that random practice produces better carry-over to sports.[16]

* See Resources section on page 219.

FUNCTIONAL EXERCISE

Exercise	Rest	Intensity	Reps	Tempo	Sets
Grip Trainer	L/R	#	12-20	Slow	2-6

Perform Grip Trainer exercises as outlined in Phase II.

Figure 5.38A
Grip Trainer: Finger Flexors

Figure 5.38B
Grip Trainer: Finger Extensors

CHAPTER FIVE

PROGRAMMING OPTION A DAILY PLANNING WORKSHEET

TRAINING PHASE: Phase III - Dynamic Stability Exercises
OBJECTIVE: Increase body awareness and coordination
DATES: 4-6 weeks

Dynamic Stability - Phase III

Exercise	Rest	Intensity	Reps	Tempo	Sets
Stretch & Warm Up Set First					
Monday					
Supine Hip Extension:Knee Flexion	1:00		8-12	Slow	1-3
Tuesday					
Fitter Wobble Board*	1:00	##	1-3	30 secs	1-8
Wednesday					
Supine Lateral Ball Roll	1:00		6-8	1 sec hold	1-3
Thursday					
Prone Twister: Feet on Ball	1:00		6-10	S-M	1-3
Friday					
Kneeling on Swiss Ball	1:00		1-3	Hold 30 secs	4-8
Saturday					
Supine Russian Twist	1:00	-2 reps	6-10 each	S-M	1-3
Sunday					
Prone Jack Knife	1:00		8-12	202	1-3
Pro Fitter*	1:00		Max**	S-M-F	1-8
Grip Trainer	L/R	#	12-20	Slow	2-6

*: These are exercises best suited for young golfers or those in good physical condition. They are not recommended for senior golfers as they require significant balance and reaction time.
See next page for additional notes.

PROGRAMMING OPTION B DAILY PLANNING WORKSHEET

TRAINING PHASE: Phase III - Dynamic Stability Exercises
OBJECTIVE: Increase body awareness and coordination
DATES: 4-6 weeks

Dynamic Stability – Phase III

Exercise	Rest	Intensity	Reps	Tempo	Sets
Stretch & Warm Up Set First					
Monday					
Supine Hip Extension:Knee Flexion	⇩1:00		8-12	Slow	1-3
Prone Jack Knife	⇩⇨⇧		8-12	202	1-3
Wednesday					
Fitter Wobble Board*	⇩1:00	##	1-3	30 secs	1-8
Prone Twister: Feet on Ball	⇩⇨⇧		6-10	S-M	1-3
Friday					
Kneeling on Swiss Ball	⇩1:00		1-3	Hold 30 secs	1-8
Supine Russian Twist	⇩⇨⇧	-2 reps	6-10 each	S-M	1-3
Sunday					
Pro Fitter*	⇩1:00		Max**	S-M-F	1-8
Supine Lateral Ball Roll	⇩⇨⇧		6-8	1 sec hold	1-3
Grip Trainer	L/R	#	12-20	Slow	2-6

**: Perform only as many repetitions as possible without losing form. Stop if you begin to feel fatigued. These exercises require high levels of coordination.
##: Board Level 1, 2 or 3
#: Select the density Power Web that will put you in the correct rep range.
See previous page for more notes.

CHAPTER FIVE

PROGRAMMING OPTION C DAILY PLANNING WORKSHEET

TRAINING PHASE: Phase III - Dynamic Stability Exercises
OBJECTIVE: Increase body awareness and coordination
DATES: 4 weeks

Dynamic Stability – Phase III

Exercise	Rest	Intensity	Reps	Tempo	Sets
Stretch & Warm Up Set First					
Training Day 1 & 3					
Fitter Wobble Board*	⇩ 1:00	##	1-3	30 secs	1-8
Supine Lateral Ball Roll	⇩		6-8	1 sec hold	1-3
Prone Twister: Feet on Ball	⇩		6-10	S-M	1-3
Supine Hip Extension:Knee Flexion	⇩ ⇨ ⇧		8-12	Slow	1-3
Training Day 2 & 4					
Kneeling on Swiss Ball	⇩ 1:00		1-3	Hold 30 secs	1-8
Supine Russian Twist	⇩	-2 reps	6-10 each	S-M	1-2
Pro Fitter*	⇩		Max**	S-M-F	1-8
Prone Jack Knife	⇩ ⇨ ⇧		8-12	Slow	1-3
Grip Trainer	L/R	#	12-20	Slow	2-6

*: These are exercises best suited for young golfers or those in good physical condition. They are not recommended for senior golfers as they require significant balance and reaction time.
**: Perform only as many repetitions as possible without losing form. Stop if you begin to feel fatigued. These exercises require high levels of coordination.
##: Board Level 1, 2 or 3
#: Select the density Power Web that will put you in the correct rep range.

CHAPTER 6

STRENGTH TRAINING

It is often said that serious golfers will do anything to improve their game. At this point, you have had twelve weeks worth of valuable exercises. If you have done the twelve week flexibility-stability program, you should be feeling the benefits of improved stability and postural alignment. It is now time to learn about the many performance enhancing effects achievable with the Strength Training phase of your program. Although golfers in general avoid strength training as a means of improving their game, there are significant benefits to both game performance and spinal health gained through a good strength training program. To illustrate this point, consider the following facts:

- The number one injury among male golfers is back pain (53%) followed by elbow pain (24%).[1]
- The number one injury among female golfers is lower back pain (45%) followed by elbow pain (27%).[1]
- As many as 63% of novice golfers are reported to suffer from back pain.[2]
- The most common injuries among professional golfers are back and wrist related.
- Up to 30% of professional golfers on tour are playing injured.[3]
- Nearly 60% of those who had been injured playing golf reported they were still troubled by their injuries.[4]
- Biomechanical analysis of the golf swing reveals that the forces generated in the spine are great enough to fracture vertebra and damage lumbar disks![5]
- Golfers participating in other sporting activities are 40% more likely to develop back pain than those who just golf.[6]

One need not be a statistician to see a trend developing here: GOLFERS NEED TO BE STRONGER TO SURVIVE THE SPORT! In fact, it is reported that amateur golfers achieve approximately 90% of their peak muscle activity when driving a golf ball.[7] This degree of muscle activation puts golf right up there with sports like football, hockey and martial arts for comparable levels of exertion, the only difference being that other athletes placed under such physical demand see strength training as an integral part of their success.

When the golfer regularly performs strength training, the key musculature surrounding the spine and pelvis becomes strong enough to protect the spine, attenuating forces away from vital bony and neurological structures (Figure 6.1).

Elbow and wrist injuries, not just back injury, may be prevented with use of golf strength exercises. Yes, you read correctly; golf strength exercises, particularly those focusing on the trunk, can reduce chances of elbow and wrist injuries.

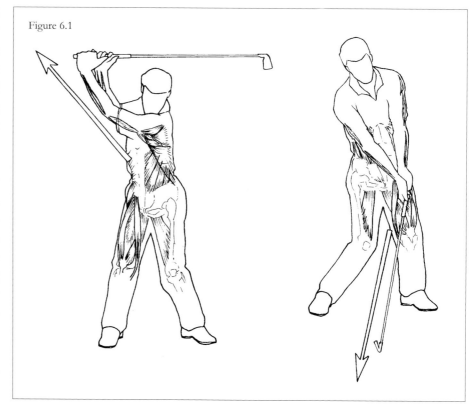

Figure 6.1

What the performance minded golfer needs to realize is that the golf swing is initiated by the abdominal muscles some 30 milliseconds before the arms or legs ever move.[8,9] The leg and arm muscles are recruited from the core outward. It is only through activation of the abdominal muscles, followed by the leg and arm muscles, that the spine and pelvis are functionally stabilized enough to allow a powerful swing without traumatizing the spinal structures. If the core musculature is weak or does not work properly, the golfer will inevitably overuse the arms to compensate, resulting in micro-trauma to the muscles and tendons.

Some common examples of swing faults that result in overuse of the wrist flexors and extensors are:

- Casting the clubhead
- Under-rotating the pelvis and/or shoulder girdle
- Poor weight transfer
- Reduced extension in the thoracic and/or lumbar spine
- Poor position at impact
- Swinging out-to-in

STRENGTH TRAINING

Not surprisingly, a common orthopedic injury has been named after golfers. The term "Golfer's Elbow" describes pain at the medial elbow where the wrist flexor tendons insert. Overcompensating for weak trunk musculature with the wrists can cause this problem.

Whole in One Golf strength training also assists the golfer in getting around the course without fatiguing to levels that may impede performance. This is very important because of the well-known fact that as one fatigues, the level of fine motor coordination diminishes.[10] What this means to you is that if you tend to play better in the first half of the game than the second half, you are probably suffering from the effects of fatigue!

Another very important benefit of strength training for golf is that it provides the training stimulus to get stronger. Strength is a key component of power. It is power that dictates just how far you can drive the ball. When you combine the components of this program, you have the foundation for huge improvements in your golf game.

Although this relatively complex topic has been condensed, it should be clear that golfers *are athletes* and based on injury statistics, should train for the game. If you perform the exercises outlined in the Golf Strength section for four to eight weeks, you will be ready to move on to the Golf Power exercises. The Golf Power exercises are designed to improve your explosiveness and dynamic strength, enabling you to drive longer than ever before! Don't miss out on the chance to capitalize on the Golf Power experience. Do your strength training exercises!

Strength is a key component of power. It is power that dictates just how far you can drive the ball.

CHAPTER SIX

GOLF STRENGTH – PHASE IV

You have now come to the point where progression will require a more significant time contribution. Several of the exercises presented in the remainder of your conditioning program will require that you either have a gym at home, or go to a local gym that is properly equipped.

Phases IV and V of your conditioning program represent an important developmental progression for most golfers. Many of you will have already made significant improvements in your golf performance. Strength training provides a foundation that is not only important for golf performance, but also for spinal health. Up to this point, your conditioning programs have been designed to give you the needed working foundation to benefit from strength and power training without increasing your risk of injury on the golf course. To maximize your results, continue through the program.

Some of you may be saying, "Hey, I want to go all the way!" Some may be saying, "I want to go all the way, but I don't want to go to the gym and be around bodybuilders to finish my program." That is fine, there is an alternative. You can purchase a Total Gym or Gravity machine (see Resources section). These are two of the best and most versatile pieces of home exercise equipment on the market today.

With the use of a Total Gym or Gravity machine, you or your client can reproduce many of the exercises suggested in the remainder of your golf conditioning program in the comfort of your own home. The only additional tools you will need, above the items already mentioned in Chapters 4 and 5, are a few medicine balls. These are weighted balls ranging in size from a baseball to larger than a basketball and in weight from $1/2$ lb. to 20 lbs. They can be used in place of dumbbells or as a training tool in their own right, as you can throw and catch them (unlike a dumbbell!).

Completing Phase IV will require that you perform Programs A and B at least one time every four days. For example, if you start with Program A on Monday, you will need to repeat it again by Friday. Between Monday and Friday, you will need to have completed Program B one time as well.

This training schedule is not based on a seven day week. You simply make sure you get your exercise programs done one time every four days. If you play golf on the same day that you are scheduled to do your exercises, always play golf first! If you try to golf after doing your exercises, your body will already be fatigued. This will most likely result in poor performance on the golf course.

Optimizing Your Work / Rest Ratio

Everyone has different physiological responses to resistance training. Some people can handle large amounts of volume, while others can handle high intensities but prefer less volume. Overtraining can be a concern.

Overtraining results in reduced performance and frequently results in diminished drive and desire to continue an exercise program. To prevent overtraining in Phase IV and V of your golf conditioning program, it is suggested that you cut the volume in half at the start of and through the fourth week of your program. This will allow your body time to recover and you will feel much stronger at the beginning of the next phase, or for the remainder of your current phase, should you opt to extend your phase to six weeks.

Phase IV Equipment Needs

Phase IV should be completed at a professional, fully equipped gym, or at home with the use of a Total Gym or Gravity machine. At this point, there should be no phobia about exercising with other people in a gym, as you will be surprised to find that having completed Phases 1-3 will have given you ample strength and coordination to feel at home among the Gymmies!

Just remember, "*Golfers are athletes!*" If you want to achieve your golf potential, you must continue the journey. When you are looking for a gym, request the following equipment for the completion of Phase IV (see Phase V-VII for other gym needs):

- Adjustable cable cross machine, or Total Gym or Gravity Machine
- Wooden dowel rod 6' x 1⅜" diameter
- Box Step
- Olympic Bar and weight plates
- Blood pressure cuff with extender hose*
- Adjustable training blocks for performing Dead Lifts

* See Resources section on page 219

CHAPTER SIX

PHASE IV EXERCISES - PROGRAM A

Exercise	Rest	Intensity	Reps	Tempo	Sets
Standing Single Arm Cable Push	⇩+1:30	-2 reps	8-12	202	1-3

An important part of golf is being able to effectively integrate your arms, torso and legs. If you have strong abdominals from doing crunches on the floor or machines, strong legs from the knee extension or leg press machines and strong arms from the bench press, it is unlikely that your golf performance will improve! In a golf situation, your brain cannot use these isolated pathways because the circuitry has not been put in place to integrate the muscles into an organized "whole."

The Standing Single Arm Cable Push exercise will teach the brain how to:

- Stabilize the body against an external load

- Integrate the legs, torso and arms

- Properly recruit the core musculature to stabilize the pelvis and shoulder girdle, just as you will need to do during a golf swing!

- Adjust the cable system so that the cable pulley is level with or slightly above your shoulder (Figure 6.2).

- Assume a stable stance with the cable in one hand and the opposite leg forward. This is called a contra-lateral split stance. There should be enough room to allow your trunk to rotate backward and not bump your elbow on the machine.

- Keep your knees soft (unlocked) or slightly bent.

- Hold the elbow in the same horizontal plane as the shoulder throughout all repetitions.

- From the start position (see Figure 6.2A), initiate the movement by drawing the navel toward the spine and then rotating the trunk.

- Once the navel is drawn in and the trunk initiates rotation, integrate the shoulder into the movement, pushing the cable handle forward (Figure 6.2B). It is very important to think of generating a movement that emanates from the core outward. The most common mistake made is to "push with the arm." Remember, the arm is just an extension of the trunk!

The exercise should be performed at a tempo of 202 for 8-12 repetitions each side. The load used for your training sets should allow completion of 8-12 reps, yet you should always stop with the feeling that you could do two more repetitions with perfect form.

After completing your repetitions on one side, switch to the other side and then progress immediately to the next exercise. You will notice when looking at your Daily Planning Worksheet that this is a circuit style program (indicated by the arrows in the rest column). When performing a circuit program, complete the entire circuit at 60% intensity as a warm-up. The warm-up circuit will not be considered part of your work sets. You should still complete between 1-3 circuits after the warm-up.

If you feel comfortable doing more than one circuit to start, you may. Just be careful not to overdo it! You will know it is time to add another circuit when you have no post-exercise soreness at the time of your next Program A training session.

Figure 6.2
Standing Single Arm
Cable Push

CHAPTER SIX

Exercise	Rest	Intensity	Reps	Tempo	Sets
Multi-Directional Lunge	⇩		1-3 each	Moderate	1-3

The Multi-Directional Lunge exercise is very beneficial to the golfer, as it carries over to several aspects of the game. Golf requires a fair bit of walking, which is a series of mini lunges. There are times when you must address the ball in various types of terrain. Standing in the rough on the side of a hill requires movement skills that are developed by the Multi-Directional Lunge.

Performing the Multi-Directional Lunge (MDL) requires activation of all the muscles surrounding the hip joint. This is important to the golfer because keeping these muscles fit prevents injury to the hips and pelvis. The MDL also trains the nervous system to be able to move in many movement patterns that you may encounter in the course rough, or in everyday life. Good postural alignment, balance and coordination are all trained when performing the MDL. These are all biomotor abilities that are either undeveloped or poorly developed on machines in the traditional gym setting.

- Place a wooden dowel rod (or light weight bar) across your back.

- Grip the bar as close to your body as you comfortably can.

- Take a deep, diaphragmatic breath (full belly then full chest).

- Without letting your breath out, draw your navel inward slightly toward your spine, as though you were sucking your stomach in to tuck in your shirt. This will activate the transversus abdominis muscle, which will in turn activate the thoraco-lumbar fascia mechanism (a key stabilizer system in the body).

- Hold an upright posture with breath and stomach drawn in. Keep hips square to the front.

- Step forward into the lunge with your left leg (Figure 6.3A). If your step length is correct, your front shin will be vertical.

- Allow your body to descend into the lunge as deeply as possible, or until the trailing knee is just off the floor.

- Release the air through pursed lips as you return to the start position, pushing back with the front leg. Don't let the air just escape unrestricted.

Figure 6.3A
Forward Lunge

STRENGTH TRAINING

Initially, the golfer should be very careful not to come out of the lunge position too quickly. A moderate, or 202, tempo is a comfortable speed of movement, not slow, not fast. If you have the urge to bounce out of the lunge position, it is better to half-step the return, which requires less force and is not hard on the knees.

Keep mild pressure in the lips so you have to keep the abdominal muscles activated to help expel the air. This will protect you from placing too much pressure on your heart while protecting your spine from being unsupported by the stabilizer mechanisms of the back. (For more information on stabilizer mechanisms, see Paul Chek's DVD course, *Scientific Back Training*.)

Having returned to the start position, you will now prepare to perform the forward lunge at an angle of 45° to the left of your first lunge (Figure 6.3B). While performing this lunge, it is very important to keep your head and eyes forward, shoulders and pelvis square to the front and allow the trailing leg to pivot naturally as you drop into the lunge. A common mistake is to turn the whole body 45° and lunge, which is no different than a front lunge.

Figure 6.3B
Forward 45° Lunge

You will feel that there are different muscles being exercised around the hip as you change angles each time. If you don't feel the change in muscles being worked, you are most likely performing the exercise incorrectly.

- Lunge forward at a 45° angle.

- Do not allow the back heel to drop inward. This will place unwanted stress and torque on the knee joint. The knee and ankle joints are hinge joints and should not be unnecessarily torqued during training exercises.

- Again, return to the start position by either a single or double step method.

Next is the lateral lunge. The lateral lunge pattern is often challenging for golfers, as are the backward lunges, which is all the more reason to master the exercise. You don't want to wait until golf forces you into one of these positions, be unable to react correctly and tear a muscle, do you?

Figure 6.3C
Lateral Lunge

- Step laterally into the lunge as shown in Figure 6.3C.

- Keep your feet pointing forward or just slightly turned out.

- Keep your torso upright and your head and eyes positioned so that you are looking across the horizon.

- Drop only as far into the lunge as you can with perfect form.

The backward 45° lunge position is the next part of the lunge exercise (Figure 6.3D).

- Look backward to get an idea of which direction you are stepping toward. This is helpful since many people's bodies will avoid the 45° pattern because it is foreign to them and requires that the brain orchestrate a new movement.

- The breathing, transversus abdominis activation and knee and ankle action are all the same as with the forward lunge, you are just going backward.

The backward-center lunge is next (Figure 6.3E). It is performed in exactly the same way as the front lunge, with the same procedure, but step backward.

Figure 6.3D
Backward 45° Lunge

Figure 6.3E
Backward Lunge

If you find some of these lunges challenging, that is very good. This means your nervous system is learning from the experience. If you want to get something from an exercise, then you have to challenge your body.

STRENGTH TRAINING

Putting the Lunge Sequence Together

For beginners with low levels of initial conditioning, alternate from the left to the right leg with each lunge. This way one leg rests while the other works. When your strength levels rise, and you feel you can take a bigger workload, perform all the lunges on one leg and then switch to the other.

Special Precautions

Many of you will no doubt have concerns regarding degenerative knee or hip joints. The Multi-Directional Lunge exercise can be modified for such cases in the following ways:

- Slow the tempo down to 313, which requires a dead-stop pause at the bottom of the lunge. Slower speeds of movement require that you work the muscle longer during each set and that you have less assistance from inertia. Therefore, you can get a good workout with very light or no external load.

- Reduce the number of repetitions performed to one or two per position, which is adequate when combined with slower tempo training.

- Move only through a pain-free range of motion. There is no benefit gained by training in pain!

- Perform each rep in a split squat format, which requires that you don't step out of the lunge until all repetitions are completed. For example, you would perform your front lunge to begin, stay in that position rising up and down two times and then return to the start position. This will reduce the wear on joints significantly.

CHAPTER SIX

Exercise	Rest	Intensity	Reps	Tempo	Sets
Wood Chop	⇩	-2 reps	10 each	202	1-2

The Wood Chop exercise is one of the most effective and universally applicable exercises. The Wood Chop is named after the pattern of movement created when chopping wood with an ax. This flexion/rotation pattern of movement is a catalyst to successful performance in many sports, for example:

- Playing baseball
- Playing hockey
- Throwing a football
- Swinging a golf club
- Passing a basketball
- Tossing a rugby ball
- Throwing a javelin
- Performing a throw in judo

These are just a few sporting examples for which the Wood Chop pattern has relevance in training for success. The Wood Chop pattern is an integral part of the "twist pattern." The twist pattern is one of the key movements or generalized motor patterns from which the brain makes other similar patterns. You can learn more about generalized motor patterns by studying the work of Richard Schmidt.[11]

With strict regard to golf, the Wood Chop exercise provides the strength foundation for developing power for driving and middle distance shots. Club head speed is generated from the top of the backswing to impact, and it is in this exact range of movement and combination of movement planes that the Wood Chop is dominant. No crunch device or abdominal machine in the world can provide the performance gains of the Wood Chop exercise! Anyone trying to convince you otherwise has something to sell.

Initially, the Wood Chop should be performed with a stable pelvis and base of support. This will allow more effective isolation of the trunk rotator muscles, which is important for the golfer with less than optimal trunk strength. When the trunk rotators are insufficient, the brain cheats the movement by over-using the legs and arms. This is one of the major reasons for such high incidence of wrist and elbow injuries in golf.[12]

You will perform this exercise with a crossover machine or a cable column. With the cable pulley positioned at the top of a crossover machine, reach up and grab the handle with the hand opposite the cable column (Figure 6.4A). If you rotate your trunk to the right, you would first grab the handle with your right hand. Then place the left hand on top of the right hand (in this case).

- Take a stance with your feet shoulder-width apart, or slightly wider.
- Initiate the rotation of the trunk by drawing your navel toward your spine.
- Think of expressing the rotational movement from the trunk outward rather than starting with the arms. If you think of pulling on the handle with your hands and arms, your recruitment pattern will be inverted and inefficient, increasing the chance of shoulder injury, just as if you initiate a golf swing with your arms.
- Pull the handle downward and across the body, as though you were taking the handle to your pocket.

- End the movement when your hands reach just outside your right pocket.

In the first stage of learning the Wood Chop, you should not allow your pelvis to deviate laterally as you perform the exercise. Be sure not to bend forward any more than necessary to get your hands to the point just outside your pocket and to the lateral aspect of your leg.

The exercise should be performed with a 202 tempo for ten repetitions. As you can see in the intensity column, there is a two repetition buffer (-2 Reps). This indicates that you should choose a weight that allows you to complete about twelve reps feeling as though you could perform two more good reps. This is a way of ensuring that you don't overload yourself and degrade the quality of the movement pattern being recorded by the nervous system.

Wood Chop Stage II

After having performed the Wood Chop exercise with a fixed pelvis for 3-4 training sessions, you can progress to performing the exercise with dynamic lower body action. Think, for example, of how you move when swinging a golf club; you move your trunk, legs and arms in synergy, not in isolation.

In Figure 6.4B, you can see the dynamic nature of the WC when performed with a weight shift. When the exercise is performed with upper/lower body integration, you should end up with 70% of your weight on the outside leg and 30% on the inside leg.

It is very important to initiate the exercise by drawing the navel inward. It is also very important to continue to think of the arms acting as an expression of the trunk. This high quality training will carry over to the place you want to see the results the most, *the golf course!*

> You must always remember the golden rule of corrective and performance exercise:
> **Junk in = Junk Out!**

Figure 6.4
Wood Chop with Weight Shift

CHAPTER SIX

Exercise	Rest	Intensity	Reps	Tempo	Sets
Box Step-Up	⇩	-2 reps	10 each	Moderate	1-3

The Box Step-Up is not only an excellent general strength and conditioning exercise, it is very useful for re-establishing normal function of the hip, knee and ankle stabilizers.

Ideally perform the exercise with no shoes on. Place a line down the center of your kneecap and another line down the second toe. These two lines should stay lined up during the exercise.

Initially, no additional weight should be used. To determine how high the step should be, you will need to start with only 2" and progress upward. You will know when you have passed your optimal working height when you step up onto the step box and can no longer keep the line on the knee tracking over the line on the second toe.

Most people need to start at heights lower than the traditional step benches used in aerobics classes, so make sure your step is adjustable and can start with a low level platform.

- Place one foot on the step.

- With good upright posture, slowly accept your weight as you step up onto the step. It is considered poor form to allow the body to bend at the trunk or to allow the line on your kneecap to move inside the line on your second toe.

- Slowly come down to the point where the back foot touches the ground, but do not fall backward off the step! That is avoiding the weak spot, the spot in the range of motion that needs to be conditioned the most.

It is very beneficial to train in front of a mirror so you don't have to look down at your knees and toes. Looking downward will induce

Figure 6.5
Box Step-Up

STRENGTH TRAINING

forward migration of the head and loss of optimal spinal curvatures leading to unwanted developments in the nervous system. These developments are likely to show up on the golf course as poor posture at address.[13]

The Box Step-Up exercise should be performed such that you fatigue between 8-12 reps. If you find that you can do more than twelve reps, but can't go to the next step height without losing form, then stay with the current step height and hold some small dumbbells. This will strengthen you as you re-educate the nervous system and activate the hip rotator muscles that have probably been on vacation for some time. *The absolute key here is to only train with perfect form.*

When you can perform the prescribed reps for the prescribed number of sets, simply elevate the step height. This will again make the exercise more challenging and further stimulate the body to get stronger.

CHAPTER SIX

Exercise	Rest	Intensity	Reps	Tempo	Sets
Lateral Shoulder Rotators	⇩ ⇨ ⇧	-2 reps	8-12 each*	303	1-3

Two of the lateral shoulder rotators (infraspinatus and teres minor) are rotator cuff muscles and are very important to any golfer. Figure 6.6A demonstrates conditioning for the lateral rotators of the shoulder from what is called the neutral position. Figure 6.6B demonstrates conditioning from what is called the plane of the scapula. These are both necessary conditioning positions for base rotator cuff conditioning that will be expanded upon in later phases of your golf conditioning program.

The Whole in One approach is to perform corrective exercises as a prophylactic measure. These two shoulder exercises help golfers prevent future shoulder problems.

Change between these two positions each training session. This will allow a more complete strengthening of your rotator cuff in a short time.

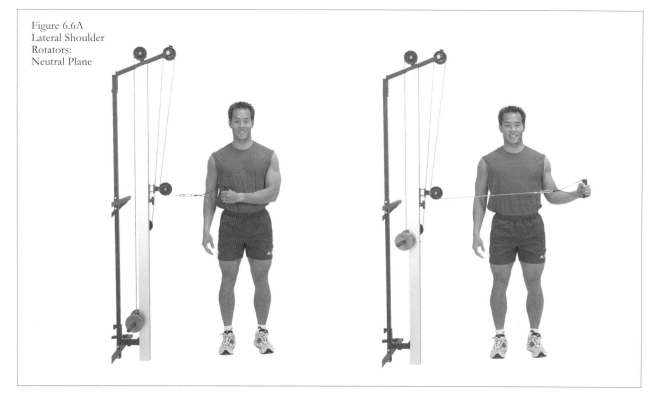

Figure 6.6A
Lateral Shoulder
Rotators:
Neutral Plane

* You will notice an asterisk in the Reps column. This is a reminder to reduce the intensity or load if there is any shoulder discomfort during the exercise. If need be, start with a load allowing 12-20 repetitions. *If for any reason you experience soreness beyond the typical post-exercise muscle soreness from doing these exercises, you should see your doctor.*

STRENGTH TRAINING

Figure 6.6B
Lateral Shoulder
Rotators: Scapula
Plane

Session One

- Use the neutral position (Figure 6.6A). This position requires that you keep the humerus (upper arm) stable at your side, with the elbow bent at 90°.

- Grab the handle of the cable pulley, adjusted so that the cable is parallel to the floor.

- Starting with your forearm close your body, pull the cable across in front of you.

- The upper arm or elbow should not wander as you perform the exercise or you will be recruiting the deltoid muscle more than the small external rotators. One useful technique to assure proper arm placement is to put a small medicine ball between the upper arm and body as you do the exercise. Having to hold the ball in place helps keep a consistent axis of rotation.

Session Two

- Position the arm about 30-45° forward of the midline of the body from the lateral perspective (see Figure 6.6B) and raised to the side 60°-80° (i.e. below shoulder level).

- Grab the handle of the cable pulley, adjusted so that the pulley is around knee height.

- Starting with your palm facing the floor, pull the cable up until your forearm is vertical, using a 303 tempo.

- Again, the upper arm or elbow should not wander as you perform the exercise. Additionally, it is important to maintain good upright posture, not letting the shoulders round forward.

CHAPTER SIX

PROGRAM B

Exercise	Rest	Intensity	Reps	Tempo	Sets
Single Arm High Cable Row & Row & Reach	⇩	-1 rep	8-12	202	1-3

The Single Arm High Cable Row is performed in much the same manner as the Single Arm Cable Push in Program A.

- Stand in a split stance.

- Hold the cable handle in the hand opposite the forward leg.

- Initiate the movement from the position of a forward weight shift. There should be very little torso rotation, as the pulling action should come from a stable lower body and working shoulder arm complex.

- Just prior to initiating the pull, there should be mild contraction of the transversus abdominis, which is done by bringing the navel toward the spine.

- As with the Wood Chop and Single Arm Cable Push exercises, the arm should be thought of as a working extension of the trunk. Your thought process should not be to pull with the arm; it should be to pull with the trunk and legs and let the arm be the connecting link that lets it happen.

- Keep the forearm in exact alignment with the cable as the handle is drawn back. At no point during the exercise should your elbow drop down below the line of pull, nor rise above it. Done incorrectly, this movement will encourage wrist flexor and extensor tendonitis!

Figure 6.7A
Single Arm High Cable Row

Figure 6.7B
Single Arm High Cable Row and Reach

The exercise should be performed on one side and then the other. The load should allow the completion of about 12 reps on the first set, stopping when you feel you have one more good rep left. This buffer zone allows you to keep the reps at or above eight in the later sets. After completing the exercise on the left and right sides, progress to the next exercise in the circuit.

At the halfway point of Phase IV, you should progress the exercise to the Single Arm High Cable Row and Reach, which is a more fully integrated version of the exercise. The Single Arm High Cable Row and Reach encompasses a reach and pull technique (Figure 6.7B). There will now be a significant shift toward the rear foot as you complete the pulling action and a forward weight shift as you return to the start position.

Looking at Figure 6.7B, you will see that the reaching arm reaches along the exact line of the cable. This reaching motion has some very beneficial effects as it encourages rotational forces through the middle back, maintaining proper function in the spinal joints. This is particularly important to the golfer because as these joints stiffen, drive distance decreases and chance of compensatory shoulder injury rises.

CHAPTER SIX

Exercise	Rest	Intensity	Reps	Tempo	Sets
Reverse Wood Chop	⇩	-3 reps	6-8 each	202	1-3

The Reverse Wood Chop exercise comes with all the benefits outlined for the Wood Chop. To better explain the benefits of this truly great exercise, I will quote Newton, who stated, "For every action there is an equal and opposite reaction." With this statement in mind, consider that the pitcher must first cock his arm and coil his trunk to deliver the fast ball, the hockey player must wind-up to deliver the slap shot and the golfer must execute the backswing in preface to the downswing. One cannot happen without the other.

It is very important that you have normal flexibility in the thoracic spine or this exercise could make your shoulders sore. To assess yourself, refer to the Thoracic Extension Test, Figure 2.20.

The Reverse Wood Chop exercise is one that improves your ability to coil, or to wind the body's natural spring mechanism.

- To begin the exercise, grasp the handle with the outside hand first, then place your other hand over the outside hand.

- Position your legs as shown in Figure 6.8A with 70% of your weight on your inside leg and 30% on the outside leg. Hold your chest high and look forward, not down.

Figure 6.8
Reverse Wood Chop

STRENGTH TRAINING

- It is very important not to allow the low back to round, as this places unwanted stress on the lumbar discs. To prevent losing your lumbar curvature, use the taping technique described in Chapter Three, Figure 3.17C.

- In the bottom or start position, take a deep breath and draw the navel inward.

- Extend the body while rotating upward and diagonally.

- At the top of the movement, the weight distribution should be 70% outside leg, 30% inside leg (Figure 6.8B).

If you look carefully at the acute variables listed in the table for the exercise above, you will see -3 Reps in the intensity column. This is very important since this complex exercise is done right before the Dead Lift in your circuit. If you overdo it, you will be too tired to perform the Dead Lift properly. Remember that in the beginning phases of your conditioning program, the majority of your strength gains are neural, or nervous system adaptations. This means that the primary mechanism by which you get stronger is learning.

CHAPTER SIX

Exercise	Rest	Intensity	Reps	Tempo	Sets
Dead Lift off Blocks	⇩	-2 reps	10 each	202	1-3

The Dead Lift exercise is considered a staple exercise by many in the strength and conditioning profession. The exercise closely resembles one of the seven key Primal Pattern® Movements, the bend pattern.[14]

The bend pattern is particularly relevant to golf because every time you bend over to address the ball, or to pick it up, you are using this pattern. The Dead Lift (DL) exercise is particularly useful for golfers for the following reasons:

- Golfers have a very high percentage of back pain.[15,16,17,18]
- The DL exercise teaches the golfer how to bend correctly and integrate the trunk and legs.
- The DL provides improved dynamic postural strength for the golfer.
- The DL aids in restoring normal flexibility in the bend pattern.
- Achieving optimal movement skill through use of the DL exercise will have benefits reaching far beyond the golf course.

The Dead Lift is regularly performed from the floor, but for this program it is modified slightly. Many people who attempt the Dead Lift from the floor don't have the flexibility or motor skill to perform the exercise correctly. To perform this crucial exercise with correct form, start by performing it from blocks, which reduce the range of motion to the point that even the beginner can perform with good form. (See Resources section)

Position the box height such that it allows you to maintain a natural curvature in the low back at the lowest point in the lift.

- Stand with your feet hip-width apart and your shins against the bar.
- Grip the bar just outside the knees with the elbows turned back so that the biceps muscles face forward as much as possible.
- Hold your chest high with the shoulder blades together.
- Hold your head so that it is in neutral alignment with the spine.
- Look forward and be careful not to tip the head upward too much.
- Take a deep diaphragmatic breath, hold and pull the navel toward the spine to activate the stabilizer mechanism of the body.

- Lift the load from the blocks with the mental image of "pushing the earth away," not pulling the load with the arms. This is critical because initiating the lift from the arms encourages overuse of the back, underuse of the legs and constitutes poor form.

- From the point of lifting the weight to the point where the bar passes the knees, the golfer should hold the trunk rigid and push with the legs.

- As the bar passes the knees, integrate the back and hips into the lift, finishing with the chest high and the shoulders slightly back, but not forced back.

- As you pass through the hardest point in the lift, allow air to begin escaping through pursed lips. This is important because it maintains pressure in the abdominal cavity and keeps the abdominal musculature active. It is also important because studies of weight lifters have shown that they get enlarged hearts from holding their breath while lifting heavy weights; a bigger heart is not necessarily better!

- Prior to lowering the bar to the blocks, the breathing process should be completed again: inhaling, holding and pulling the navel toward the spine. This may seem awkward in the beginning, but it will become second nature after a little training.

Figure 6.9
Dead Lift Off Blocks

!! You may be wondering if you need to use this breathing technique for all lifts. If the object you're picking up is heavy enough to naturally stop your breathing, then you should use this preparatory technique. If it is light enough that you could lift it at least twenty times, then you can breath naturally.

It is important to conclude your sets with enough energy to perform one more perfect repetiton. Using a 202 tempo, you should be lifting a load that allows as close to ten repetitions as possible.

Note: If you find that you have a hard time keeping your lumbar spine curved forward, have your back taped as in Figure 3.17C. Taping your lumbar spine from the bottom rib to the sacrum while in a natural standing posture (provided you don't have a flat back) will help train you to keep good spinal position. When your back is taped, if you round your back during the lift, you will feel the tape pull on your skin. If you have a hairy back, you will learn fast!

CHAPTER SIX

Exercise	Rest	Intensity	Reps	Tempo	Sets
Cross Body Tricep Extension	⇩	-1 rep	8-12 each	202	1-3

The Cross Body Tricep Extension is particularly useful for strengthening the arm muscles most active during the initiation of the downswing and during the follow-through phase after impact.

- Set the cable column about chest high (Figure 6.10).
- Hold the handle in the hand opposite the cable column.
- Position your hand so that the palm is facing the chest and the handle is positioned just below the collar.
- Pull the arm across the body, moving simultaneously from the shoulder and elbow joints.
- Keep the trunk still to aid in placing more demand on the shoulder and triceps musculature.

The tempo of the exercise is moderate at 202. After exercising one arm, immediately switch to the other. After exercising the right and left sides, progress to the Medial Shoulder Rotation exercise.

Figure 6.10
Cross Body Tricep Extension

STRENGTH TRAINING

Exercise	Rest	Intensity	Reps	Tempo	Sets
Medial Shoulder Rotation	⇩	-1 rep	8-12 each	303	1-3

Medial Shoulder Rotation exercises the large rotator cuff muscle called the subscapularis, as well as the pectoral and the anterior deltoid muscles. Conditioning the subscapularis and associated medial rotators from the neutral position is important, although activation of the subscapularis is greatest when the arm is abducted to about 90°. As mentioned for the Lateral Shoulder Rotation exercise in Program A, you should alternate between performing the exercise in the neutral and shoulder-abducted positions.

- Adjust the cable pulley to a position level with the forearm.

- Stand far enough from the cable column so that when the arm is fully externally rotated the weights do not touch the stack (Figure 6.11A).

- Keep tension on the arm.

- From the position of external rotation (see Figure 6.11A, start position), medially rotate the shoulder without moving the upper arm in any way other than pure rotation. If you have a hard time, hold a small medicine ball between your ribs and your arm while doing the exercise.

Figure 6.11A
Medial Shoulder Rotation: Neutral Position

CHAPTER SIX

- The tempo is slow with a -1 rep buffer zone to ensure good form.

The next time you perform Program B, you should perform the Medial Shoulder Rotations from the position shown in Figure 6.11B.

- Raise the arm to your side until it is about 10-15° below horizontal and bring it forward 30-45°.

- Perform your repetitions from this position following the same variables as described previously.

Figure 6.11B
Medial Shoulder Rotation: Arm Abducted 90°

Exercise	Rest	Intensity	Reps	Tempo	Sets
Lower Abdominal #2-B Standing	⇩ ⇨ ⇧	-2 reps	12-15	S-M-F	2-3

Lower Abdominal #2-B is performed as outlined in Phase I (see Chapter Five).

STRENGTH TRAINING

DAILY PLANNING WORKSHEET

TRAINING PHASE: Phase IV – Strength Training for Golf
OBJECTIVE: Increase functional strength
DATES: 4-6 weeks

Golf Strength – Phase IV

Exercise	Rest	Intensity	Reps	Tempo	Sets
Stretch & Warm Up Set First					
Program A					
Standing Single Arm Cable Push	⇩+1:30	-2 reps	8-12 each	202	1-3
Multi-Directional Lunge	⇩		1-3 each	Moderate	1-3
Wood Chop	⇩	-2 reps	10 each	202	1-2
Box Step-Up	⇩	-2 reps	10 each	Moderate	1-3
Lateral Shoulder Rotators	⇩⇨⇧	-2 reps	8-12*	303	1-3
Program B					
Single Arm High Cable Row π or Row & Reach	⇩+1:30	-1 rep	8-12	202	1-3 each
Reverse Wood Chop	⇩	-3 reps	6-8 each	202	1-3
Dead Lift off Blocks	⇩	-2 reps	10	202	1-3
Cross Body Tricep Extension	⇩	-1 rep	8-12 each	202	1-3 each
Medial Shoulder Rotation	⇩	-1 rep	8-12 each	303	1-3 each
Lower Abdominal #2-B Standing ψ	⇩⇨⇧	-2 reps	12-15 each	S-M-F	2-3 each

π: Keep pulling elbow elevated
ψ: Leaning against doorway or similar stabilizing object
*: Reduce the intensity or load if there is any shoulder discomfort during the exercise.

GOLF STRENGTH – PHASE V

With four to six weeks of Phase IV under your belt, you are now ready for Phase V. You will notice that the exercises in Phase V get a bit more complex, as well as more specific to golf. It is very important that you have completed Phase IV prior to beginning Phase V because there are no rewards for skipping or shortening phases!

Phase V will progress your stabilizer strengthening and conditioning as well as advance your motor capabilities and strength levels. This Phase will require that you continue in a professionally equipped gym, be it a good home gym or a membership gym.

Phase V Equipment Needs

- Swiss ball*
- Step-up box or bench*
- Cable column machine*
- Dumbbell selection
- Dead Lift blocks*
- Bath towel
- Good music!

* See Resources section on page 219.

STRENGTH TRAINING

PHASE V EXERCISES - PROGRAM A

Exercise	Rest	Intensity	Reps	Tempo	Sets
Supine Single Arm Swiss Ball Dumbbell Press	⇩	-2 reps	8-10 each	102	2-4

The Supine Single Arm Swiss Ball Dumbbell Press is an exercise that will train your body to be fully integrated, well balanced and functionally strong, helping to make you a *golf machine!*

Try using a larger size Swiss ball for this exercise and inflate it so that it is a bit softer than normal. The ball will act as a range of motion limiter for the shoulder, protecting the shoulder joint from unnecessary wear and tear. This concern grows from having to rehabilitate many shoulders in those diehard bench pressers who insist on getting the bar to the chest, a position which stretches the ligaments of most people's shoulders, leading to chronic shoulder injury.

To safely get on the ball for this exercise, choose a small dumbbell to start with. As you become familiar with the exercise, you can progressively use larger dumbbells until you find the correct weight to train within the variables outlined above.

- Sit on the ball with the dumbbell resting on your thigh.

- Roll slowly down the ball until your back is positioned such that the head and shoulder blades are supported by the ball. Bring the dumbbell to the outside of your shoulder.

- Hold your hips up so that they are as close as possible to being in the same horizontal plane as the shoulders.

- Press the dumbbell straight in the air, allowing the body to follow its natural path of motion.

- When you have completed the prescribed reps at the prescribed tempo, switch to the opposite arm and complete the set.

Figure 6.12
Supine Single Arm Swiss Ball Dumbell Press

Note: The head should not leave the ball while performing this exercise.

CHAPTER SIX

Exercise	Rest	Intensity	Reps	Tempo	Sets
Cross Box Step-up with Dumbbells	⇩	-2 reps	10 each	Moderate	2-4

If you have ever had to address a ball on the side of a hill, or had to hunt around to find your ball in the rough, this exercise will make those situations a bit easier to handle.

The Cross Box Step-Up is an excellent conditioning exercise picked up from strength and conditioning coach Loren Goldenberg. He used it for many years to condition professional hockey players, but athletes from all sports have had great success with it as well.

This exercise is an advancement on the traditional Box Step-Up performed in Phase IV and should not be performed until you have mastered that exercise. If you have achieved the ability to keep the line on the knee from crossing inside the line on the second toe at a box height of 12" or more, then you are ready to begin the Cross Box Step-Up.

Figure 6.13
Cross Box Step-Up with Dumbbells

- Place your outside foot flat on the outer edge of the box (Figure 6.13). If you are in the correct position, the angle of the shin of your leg on the box should never be greater than 30° or the angle at which you can keep your foot flat on the box surface.

- Keep an upright posture, shoulders back and chin tucked in, as though you were trying to make yourself as tall as possible.

- Take a breath in and then draw your navel toward your spine to activate the deep abdominal wall and the stabilizer system.

- Step up onto the box. As you can see in the diagram, you should step up, across the box, and then step down with the leg that you stepped up with.

- Repeat the process going in the other direction.

Choose a load that allows completion of ten repetitions (five one way and five the other), and still leaves you with enough energy to complete two more good reps. The goal is to train, not drain!

STRENGTH TRAINING

Exercise	Rest	Intensity	Reps	Tempo	Sets
Single Arm High Cable Row & Reach	⇩	-2 reps	8-12	102	2-4

Progressed from Phase IV (Figure 6.7 B), the exercise is now performed faster. In this phase you will use a 102 tempo. This equates to a quick pull and a one-two return. This technique is used as preparation for the power excrcises to come in Phases VI and VII. Do your homework, the exercises get more and more fun as we go!

Exercise	Rest	Intensity	Reps	Tempo	Sets
Swiss Ball Supine Hip Extension	⇩		8-12	303	2-3

The Swiss Ball Supine Hip Extension will further enhance your balance and strength. This exercise can be performed with a weight plate held across the lower abdominal region if you find that you are able to do more repetitions than prescribed at the suggested tempo.

- Sit on the Swiss ball.

- Roll downward until your shoulders and head are comfortably supported (Figure 6.14A).

- Extend the hips upward by pushing through your heels, always making sure to keep the shins vertical.

- Lower your butt toward the floor, again keeping the shins vertical. This will encourage optimal recruitment of your glutes.

The single leg version of the exercise requires a significant increase in balance and coordination! If your goal is to beat Tiger Woods, then you will want to go all the way. If you just want to get the best out of your body without getting too tricky, then the two-leg version will provide adequate development.

Figure 6.14A
Swiss Ball Supine Hip Extension

Figure 6.14B
Single Leg Swiss Ball Supine Hip Extension

Exercise	Rest	Intensity	Reps	Tempo	Sets
Single Arm C.R.A.C. Press	⇩⇨⇧	-1-2 reps	8-10 each	303	1-3

C.R.A.C. (Core Recruitment Antagonist Co-Contraction) exercises use functional positions with asymmetrical loads. This technique was developed specifically for improving the functional capacity of injured workers and athletes. Like most exercises developed in the rehabilitation environment, this exercise is now widely used for advanced athletic conditioning as well.

- Choose a light dumbbell to start.

- Stand facing the junction between two mirrors. If you can't find two adjoining mirrors, draw a vertical line down the center of your mirror or hang a plumb line from the center of the mirror.

- Align your body such that your nose and navel are right on the line before you attempt to press the dumbbell overhead.

- Take a deep diaphragmatic breath, hold it and draw your navel toward the spine to activate your core stabilizers.

- Press the dumbbell up, using a slow 303 tempo. As you perform your repetitions, never allow your nose or navel to shift off the line.

- Release your breath through pursed lips as you lower the weight. Reset your breathing between reps.

If you are doing the exercise correctly, you will most likely feel the fatigue in the oblique abdominals before your shoulder gets tired. Watch your pelvis closely in the mirror as you do the exercise. It is common to see the pelvis rotate or shift as the stabilizer muscles of the trunk and pelvis fatigue. If you see this happening, you must consciously correct yourself. If you are too tired to make the correction, then either switch to the other arm or end the set. After performing the C.R.A.C. Press, return to the beginning of the circuit and rest for 1:30 before starting the next circuit.

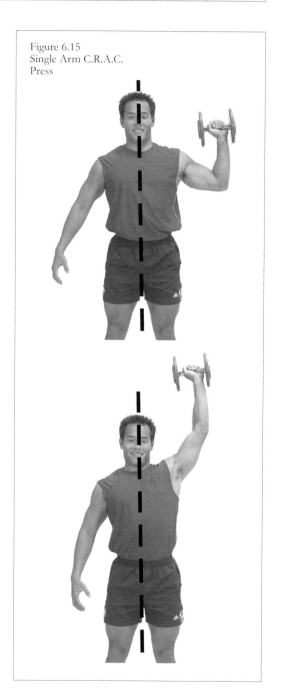

Figure 6.15
Single Arm C.R.A.C. Press

STRENGTH TRAINING

Exercise	Rest	Intensity	Reps	Tempo	Sets
Cross Body External Shoulder Rotation	L/R	-2 reps	8-10 each	102	2-3

This exercise is an advancement on the External Shoulder Rotation exercise as performed in Phase IV of strength training.

- Position the cable pulley at the bottom of the column as seen in Figure 6.16.

- Hold the handle in the hand opposite the cable column.

- The exercise begins with the arm positioned down across the body with the opposite hand resting on the opposite side pocket with the palm facing the leg.

- Bring the arm across the body to a position about 45° above horizontal, with the palm facing forward. It is critical that you externally rotate the arm before it reaches 80° of shoulder abduction (or 10° before horizontal) to protect the shoulder from unnecessary impingement.

After performing the exercise with one arm, switch to the other. Repeat this process until you have completed 2-3 sets. It is a good idea to start with only two sets and progress to the third set when you are sure there is no post-workout soreness impeding progress in your coming workout.

Figure 6.16
Cross Body External Shoulder Rotation

PROGRAM B

Exercise	Rest	Intensity	Reps	Tempo	Sets
Cross Body Triceps Extension	⇩		8-12 each	103	2-4

This exercise is performed exactly as in Phase IV, Program B. The only differences are:

- The tempo has now increased to a 103 count, which means you will bring the arm across the body in one second and return the arm over three seconds.

- You will now switch to using a towel as a handle. This is done by looping a traditional bath towel through the handle on the cable column system and holding the towel. This modification progresses your grip training.

Exercise	Rest	Intensity	Reps	Tempo	Sets
Dead Lift Off Blocks	⇩	-1 rep	8-12	303	2-3

Performed as in Phase IV, Program B, the exercise is now progressed by assessing to see if you can lower the weight further with perfect form. This is done with the use of a spotter. Ask the spotter to watch your back as you lower the weight. Whatever height you can lower the weight to without losing perfect spinal alignment will be your new height.

You will notice in the Intensity column above that you are now operating at a -1 rep intensity. This means that you should push yourself a little harder during this phase in preparation for power training in Phase VI.

STRENGTH TRAINING

Exercise	Rest	Intensity	Reps	Tempo	Sets
Cross Body Medial Shoulder Rotations	⇩	-1 rep	8-10 each	103	2-3

The Cross Body Medial Shoulder Rotation is an advancement on the Medial Shoulder Rotation exercise performed in Phase IV, Program B.

- Position the cable pulley at the top position and grasp the handle with the palm facing forward.

- Pull the cable across the body while rotating the arm medially such that the hand ends up just above the pocket on the opposite side with the palm against the hip (Figure 6.17).

Perform the exercise in one smooth motion. It should not be segmented in any way. You should complete the down motion in one second and return the arm to the top over the span of three seconds. Use a one repetition form buffer, progressing the exercise from two to three sets.

Figure 6.17
Cross Body Medial Shoulder Rotations

CHAPTER SIX

Exercise	Rest	Intensity	Reps	Tempo	Sets
Pelvis Shift with Towel Handle	⇩	-1 rep	8-10 each	103	1-3

The Pelvic Shift with Towel Handle was designed specifically to improve a golfer's ability to drive from the hips and legs. This ability is lacking in even the greatest golfers because they don't know how to train the mechanism properly.

Research shows that 54% of the force and 51% of the kinetic energy delivered to the shoulder/arm complex is generated by the lower body![19, 20] In order to drive a golf ball far, you need strong legs and a strong trunk to generate the force required in the arms to swing the golf club effectively. Again, the reason for so many wrist and elbow injuries among amateur and professional golfers is the lack of leg and core strength, resulting in arm swinging and eventual injury.

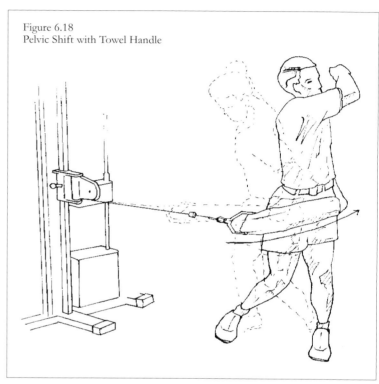

Figure 6.18
Pelvic Shift with Towel Handle

- Position the cable pulley at hip height (Figure 6.18).

- Place a bath towel through the handle, grasp it with your outside hand and assume your approach stance.

- With the arm closest to the cable column, mimic the top of the backswing; you will have to pretend to play both right and left handed to perform the exercise correctly.

- As you perform the downswing and follow-through with the free arm, use the legs and pelvis to overcome the resistance on the cable column in the exact pattern and with the same relative timing and speed you would use when swinging a golf club.

- The tempo prescribed for the exercise is 103, which means you will initiate the swing over approximately one second and return to the start position under load over three seconds.

- Make sure you stop when you feel you have one good rep left. After performing the exercise on one side, switch to the other side. Progress from one set each side to three sets as your body adapts to the exercise.

STRENGTH TRAINING

Exercise	Rest	Intensity	Reps	Tempo	Sets
Swiss Ball Side Flexion	⇩	-2 reps	6-8 each	303	1-2

The Swiss Ball Side Flexion exercise is used to improve core strength and bulletproof the golfer's back, particularly for those who carry their bags.

- Place a Swiss ball under your hip and anchor one or both of your feet along the bottom of a wall or with help of a partner (Figure 6.19).

- Hold your top leg as straight as possible and in line with the torso, shoulders and head, then lie over the ball to stretch the oblique abdominal muscles. Beginners should place their arms at their sides. Progress to placing your arms across the chest as shown, and then to having your fingertips behind the ears.

- From the bottom position (stretched over the ball), initiate the side flexion movement from the trunk.

- Side bend the spine one level at a time until the shoulders and head come up.

- On the way down, the spine should side flex over the ball one segment at a time starting from the bottom up.

- The tempo is 303, indicating that each phase of the movement should take three seconds and no hold at the top. Perform 6-8 reps, stopping when you feel you could do two more good reps. Because the oblique muscles and the quadratus lumborum (the deepest muscle in the low back) are seldom exercised this directly, start with only one set and progress carefully to the second set as your body adapts.

- Perform the exercise on each side before beginning the next exercise.

Figure 6.19
Swiss Ball Side Flexion

CHAPTER SIX

Exercise	Rest	Intensity	Reps	Tempo	Sets
Lower Abdominal #2-B Standing Unsupported	⇩⇨⇧		12-20 each	S-M-F	2-4

Lower Abdominal #2-B Unsupported is performed as demonstrated in Phase I (Figure 5.19), but without the use of the BP cuff or the wall.

- Standing unsupported, place one finger in the navel and one directly behind it on the spine.

- Stand on one leg and begin to move the free leg back and forth as if you were walking. Use your lower abdominal muscles to stabilize the pelvis and keep the spine still as you move your leg. If you are performing the exercise incorrectly, then you will feel your spine move forward and backward.

- Your navel should be drawn inward slightly to activate the transversus abdominis and stabilizer mechanism of the body.

- As you improve, increase the amplitude of the movement and progress the tempo from slow to moderate and finally to fast movements of the leg.

- Perform 12-20 repetitions on each leg per set.

STRENGTH TRAINING

DAILY PLANNING WORKSHEET

TRAINING PHASE: Phase V – Strength Training for Golf
OBJECTIVES: Increase functional strength and prepare for power training
DATES: 4-6 weeks

Golf Strength – Phase V

Exercise	Rest	Intensity	Reps	Tempo	Sets
Stretch & Warm Up Set First					
Program A					
Supine Single Arm Swiss Ball Dumbbell Press	⇩+1:30	-2 reps	8-10 each	102	2-4
Cross Box Step-Up with DBs	⇩	-2 reps	10 each	Moderate	2-4
Single Arm High Cable Row & Reach	⇩	-2 reps	8-12 each	102	2-4
Swiss Ball Supine Hip Extensioin	⇩		8-12	303	2-3
Single Arm C.R.A.C. Press	⇩⇨⇧	-1-2 reps	8-10 each	303	1-3
Cross Body External Shoulder Rotation	L/R	-2 rep	8-10 each	102	2-3 each
Program B					
Pelvic Shift with Towel Handle	⇩+1:30	-1 rep	8-10 each	103	1-3 each
Dead Lift Blocks$^\psi$	⇩	-1 rep	8-12	303	2-3
Cross Body Medial Shoulder Rotations	⇩	-1 rep	8-10 each	103	2-3
Standing Cross Body Triceps Extensions$^\pi$	⇩		8-12 each	103	2-4
Swiss Ball Side Flexion	⇩	-2 reps	6-8 each	303	1-2 each
Lower Abdominal #2-B Standing Unsupported	⇩⇨⇧		12-20 each	S/M/F	2-4

π: Use towel for hand-grip, not traditional metal handles.
ψ: Progressively lower blocks to lowest point perfect form can be maintained.

CHAPTER 7

POWER TRAINING

Only now, having developed static stability, dynamic stability and golf strength, is it safe to develop golf power. Before we go any further, I would like to explain just exactly what power is. Power is a term used to describe force, with respect to time. For example, if someone can lift 200 lbs. over his or her head, it is safe to say that he or she is very strong. A person, such as an Olympic weight lifter, often lifts 300-500 lbs. over his head, but to do so he must move the weight very quickly. It is how quickly the load is moved over a given distance that determines power, whereas strength is determined without respect for time. If you can lift a 300 lb. weight you may be strong, but not necessarily powerful.

To apply the principle of power to golf, consider how fast the club head is moving at impact. Obviously, the faster the club head is moving, the more powerful the golfer, and the greater the likelihood the ball will travel farther.

DEVELOPING GOLF POWER

As previously mentioned, the four factors that control the flight of a golf ball are:

❶ Swing Plane
❷ Clubface Alignment
❸ Angle of Attack
❹ Club Head Speed

What this tells us is that having a lot of power without being able to maintain an optimal swing plane, club face alignment and angle of attack, will only lead to looking for balls farther into the trees! You don't want to earn a reputation for having the longest hook or the longest slice on the course!

Do you want longer drives than ever before? If you said "yes," then it's time to turn on the POWER!

CHAPTER SEVEN

It is only by developing a base of static and dynamic stability and golf strength that you can expect to achieve factors 1-3 above. With the presence of these three factors comes improved accuracy and consistency, serving as a foundation for golf power. It is golf power that provides the fourth factor, club head speed.

> **To develop golf power, golfers must pay careful attention to two principles:**
> - Specificity of Exercise
> - Specific Adaptation to Imposed Demands, or the S.A.I.D. principle

As noted in the beginning of this book, it is important to pay careful attention to the selection of exercises. If you attempt to develop golf power with exercises that do not have a high carry-over to the game, you will not get the desired results.

The S.A.I.D. principle of conditioning indicates that our bodies will have a specific adaptation to demands imposed upon them by our selection of training exercises. This is exactly why body building tactics will not help the golfer improve performance.

To improve performance in the golf power phase of your conditioning program, functional exercises that continue to develop movement skill, balance, coordination and speed will be used. The exercises in your golf power program focus on development of power in the trunk, hips and rotator cuff. If you perform the exercises exactly as described here, you will significantly reduce your chances of wrist and forearm injury. As your core, hips and rotator cuff perform better and better, your need to compensate by accelerating the club with your arms will diminish.

The exercises in your golf power program will also continue to improve your timing. To be able to utilize your newly developed golf strength and power, your timing must be optimal. If your timing (synchronization of swing plane, club face angle and angle of attack) is off, your ability to generate club head speed will not serve to improve your game.

GOLF POWER – PHASE VI

You will notice on your Phase VI and VII Daily Planning Worksheets that the tempo of execution significantly increases compared to the strengthening exercises in Phases IV and V. Performing exercises at higher speeds aids the golfer in achieving higher club speed when driving the ball. A higher club speed, combined with optimal club face alignment, angle of attack and optimal swing plane results in better golf performance!

As with all previous phases of your conditioning program, carefully study the exercises prior to attempting them. Due to the higher speeds of movement required in this phase, it is very important to obtain your doctor's approval prior to performing these exercises. As speed of movement increases, so does stress on joints and connective tissues. This should not be a concern if you have successfully completed all previous stages of your development (Phases I - V).

For best results, complete each Phase VI Program once every four days. For example, perform Program A on Monday, Program B on Wednesday, Program A again on Friday, and Program B on Sunday. The next cycle of exercises will follow the same format, beginning on Tuesday, allowing a day of rest between cycles.

Some of you might experience joint soreness after your third or fourth week of power training. If this is the case, pay close attention to how many weeks into that particular phase you are when the discomfort starts. Some athletes respond better to two weeks of loading followed by one week of rest, while others respond to three weeks of loading followed by one week of rest.[1] Rest on the unloading week constitutes reducing the number of sets you do by half. If you are doing three sets, complete just two sets in your rest week and cut the reps on the second set in half.

If you are traveling or cannot perform your exercises on a scheduled day, simply start exactly where you left off, continuing the cycle on the same work/rest schedule.

The power phases are shorter than the previous phases, lasting only 3-4 weeks. This is to protect the body from overload secondary to the stress of high-speed exercises. Considering that golfers do not move heavy objects as many other athletes must, the greatest training effects are nervous system based. This means you will spend most of your time learning to move quickly, not learning to move massive loads requiring big muscles.

Carefully study the exercises in this section. Technique is of paramount importance. If you don't feel competent in your ability to carry out the suggested golf power programs, feel free to contact the C.H.E.K Institute about a consultation with a CHEK Golf Biomechanic or search the online database at www.CHEKconnect.com.

Phase VI Equipment Needs

- Wooden dowel rod (6' x $1^{3}/_{8}$") or an Olympic bar with two 10 lb. plates
- 2-6 pound medicine ball
- Gripper style medicine ball 0.5-2 lbs.*
- Swiss ball*
- Bouncing medicine balls*
- 1 full size bath towel
- 1 Total Gym Rebounder* or a partner of comparable strength

* See Resources section on page 219.

CHAPTER SEVEN

PHASE VI EXERCISES - PROGRAM A

Exercise	Rest	Intensity	Reps	Tempo	Sets
Multi-Directional Lunge	⇩		2-3 each	101	2-4

The Multi-Directional Lunge is fully explained in Phase IV of your Whole In One Golf conditioning program. In the Golf Power phase, the accelerated tempo of the exercise is designed to more closely resemble the rate at which similar movements are performed on the golf course.

Hopefully you will not need to go climbing around in the rough, but if you do, there is no better exercise to prepare you!

Exercise	Rest	Intensity	Reps	Tempo	Sets
Medicine Ball Power Swing	⇩ ⇨ ⇧	-2 reps	6-8	101	2-4

- Begin from the position of address.
- Hold a medicine ball weighing about 1-2 lbs.
- Perform the backswing motion.
- Immediately move into the downswing and continue with the follow-through.
- At the top of the follow-through, reverse the direction and perform the Power Swing as though you were a left-handed golfer (or right-handed if you are left-handed).

It is very important that you maintain perfect golf swing form. If you feel as though your form is being altered due to the weight of the medicine ball, then use a lighter ball. While performing the exercise, maintain a visual image of performing your best drive. This will help your brain sequence the recruitment of your muscles in a pattern that has a very high level of carry-over to the drive.

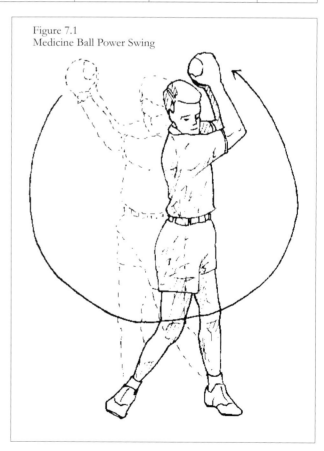
Figure 7.1
Medicine Ball Power Swing

POWER TRAINING

This exercise should be performed for 6-8 reps at a fast but controlled tempo (101). When using a correct size medicine ball, you will be able to complete 6-8 Power Swings in each direction and feel like you could do two more with perfect form. After performing the Power Swing, you should rest for two minutes before beginning your next circuit of Multi-Directional Linges and Power Swings.

As emphasized previously, you should always warm up. To warm up for the Power Swing, perform 10-12 repetitions of the exercise in each direction without a medicine ball, then do one set with half the training weight. This warm-up can be done in the exact format as the conditioning circuit prescribed with the lunges.

Exercise	Rest	Intensity	Reps	Tempo	Sets
Medicine Ball Sit-Up Pass on Swiss Ball	2:00		10-15	101	2-3

The Medicine Ball Sit-up Pass on Swiss Ball will help develop the power needed to accelerate the club head smoothly and explosively. Some of the key muscles needed to accelerate the club from the top of the backswing to impact are the abdominals, shoulder extensors and hip flexors, all of which are conditioned with this exercise.[2,3]

The exercise also serves to maintain normal range of motion in the lower and middle back as well as strengthen the neck flexors. Although the exercise can be effectively performed by throwing the ball against the Total Gym Rebounder (Figure 7.2), it can also be executed by using an air-filled medicine ball and throwing it against a wall.

- Position yourself over the Swiss ball so that your sacrum (bottom of your spine) and head both rest comfortably on the ball.

- With a medicine ball (2-6 lbs.) in your hands, initiate the toss by drawing your navel inward to stabilize the spine.

- As you sit up, throw the ball at the rebounder, being careful not to sit up so far that the Swiss ball shoots out from under you.

- Initiate the throw from your center and finish with the arms, rather than just throwing with your arms only.

Figure 7.2
Medicine Ball Sit-Up Pass on Swiss Ball

CHAPTER SEVEN

When you throw the ball, toss it hard enough so that it comes back just above your head. Catching the ball in this position will allow you to use the weight and inertial energy of the medicine ball to pull you over the Swiss ball. This will serve to pre-stretch the abdominal musculature, which will excite the nervous system and add to your power for the next toss. This is due to what is called "econcentric loading," a term developed by physical therapist Gary Gray.

Econcentric loading utilizes the muscles' spring-like qualities to generate tension. This tension serves as a spring, assisting the working muscles in throwing the ball. This is exactly what should happen at the top of your backswing. In fact, if you watch Nick Price carefully, he is very good at converting the energy of the backswing into useful power in the downswing to accelerate the club head.

Perform 10-15 reps on a fast but controlled tempo and rest for two minutes. This should be completed two to three times. This exercise will serve as a very important developmental stepping-stone to prepare you for the more ballistic exercises in Phase VII!

Exercise	Rest	Intensity	Reps	Tempo	Sets
Medicine Ball Medial Shoulder Rotators	L/R+1:00	-2 reps	8-10 each	Fast	2-4

Medicine Ball Medial Shoulder Rotations are performed from a neutral position (Figure 7.3A) and from 80°-90° of shoulder abduction (Figure 7.3B). To keep the exercises as effective as possible, focus on rotating the upper arm while maintaining a controlled movement path. In Phase IV, Program B, you performed the rotator cuff exercises with a medicine ball tucked under your arm to help focus on rotation. The same sort of isolated rotational effort should be applied when throwing the medicine ball at the rebounder.

- As you perform the medial rotation tosses, throw the ball a few times in one position and then move the elbow upward into abduction a few degrees. Repeat a few throws from several positions, progressing upward with the elbow until your arm is abducted to about 90°.

- When your arm fatigues to the point that you feel you could do two more throws before your speed and/or coordination drop, stop and begin the set on your opposite arm. After completing each arm, rest one minute before starting your next set.

If you are using the correct size medicine ball and throwing it with good form, you will reach the end of your set by the time you have performed ten repetitions. It is very important to realize that more is not better with this exercise. The goal is to train the muscle fibers responsible for high-speed contractions – the fibers that are used to stabilize your shoulder as it accelerates the club during the golf swing. If you train into fatigue, you will train yourself to be slower and hit shorter drives!

You will know it is time to use a heavier ball when you are able to perform more than ten reps on each side at high speeds with perfect form and still feel like you have two more good reps left. Increase the weight in small increments so that your speed of movement does not drop off too much as you progress.

POWER TRAINING

Figure 7.3A
Medicine Ball Medial Shoulder Rotations: Neutral Position

Figure 7.3B
Medicine Ball Medial Shoulder Rotations: Shoulder Abducted 80-90°

Exercise	Rest	Intensity	Reps	Tempo	Sets
Medicine Ball Bouncing Wood Chop	⇩		6-8 each	Fast	2-4

The Medicine Ball Bouncing Wood Chop (BWC) is excellent for developing econcentric strength. This is the type of strength needed to change directions just as you would at the top of the backswing. The BWC exercise allows you to fully express your strength as power because you actually release the medicine ball at a high speed of movement. This is not something you can do with traditional gym training. Imagine the look on people's faces if you threw a dumbbell across the gym!

Another very important benefit of the BWC is that you are able to condition your "golf muscles" in a pattern that is extremely close to an actual golf swing. When you go out to the golf course there will be a high carry-over of the power you gain from the BWC to your swing.

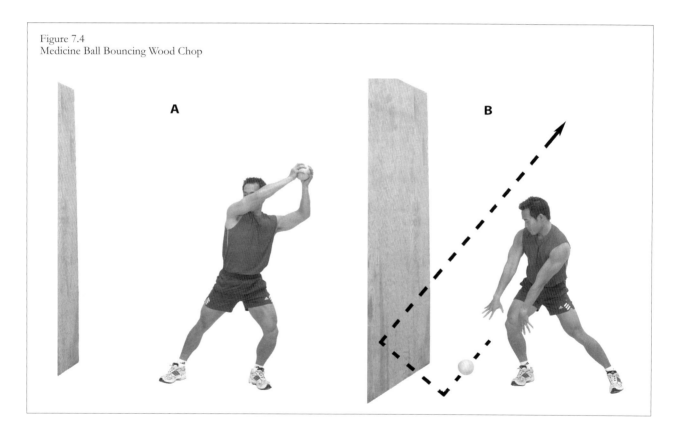

Figure 7.4
Medicine Ball Bouncing Wood Chop

- Hold a medicine ball weighing 1-6 lbs. in your hands and assume your address position.

- Coil your body into the backswing position (Figure 7.4A).

- Quickly but rhythmically move into your downswing motion.

- At the instant you change direction from backswing to downswing, you should draw the navel inward slightly to activate the deep abdominal wall and spinal stabilizers.

- Release the ball as your arms accelerate towards the floor. The ball should bounce from the floor to the wall and back to you (Figure 7.4B).

- Do the exercise on each side for a total of 6-8 repetitions per set.

Remember to always stop the exercise the instant you feel your form may begin to deteriorate. If you are unable to perform 6-8 repetitions per side, then you need to use a lighter medicine ball.

This exercise is followed directly by the Medicine Ball Reverse Wood Chop. After performing these two exercises in sequence, rest for two minutes and repeat the circuit until you have completed 2-4 sets.

CHAPTER SEVEN

Exercise	Rest	Intensity	Reps	Tempo	Sets
Medicine Ball Reverse Wood Chop	⇩⇨⇧	-1 rep	6-8 each	102	2-3

The Medicine Ball Reverse Wood Chop (RWC) is very similar in its movement characteristics to the Wood Chop. The difference is that the effort is now generated in the backswing phase. It was previously developed through the downswing phase of the movement.

- Chose a medicine ball weighing 1-6 lbs.

- The start position is very similar to that used when performing a Dead Lift with the exception that the torso is rotated just enough at the bottom position to allow the ball to be slightly outside the knee (Figure 7.5A).

- Draw the navel inward.

- Extend the trunk and shift your weight to the opposite side. The movement is diagonal and is performed the same way as the cable Reverse Wood Chop in Phase IV (Figure 7.5B).

- The movement should be fast and rhythmical for 6-8 reps on each side. After performing this exercise, rest two minutes and begin another circuit as directed in your program.

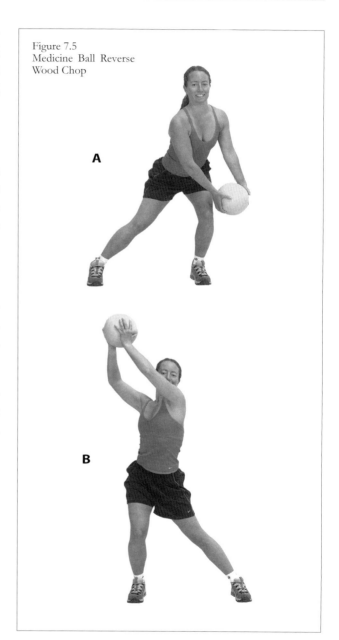

Figure 7.5
Medicine Ball Reverse Wood Chop

POWER TRAINING

Exercise	Rest	Intensity	Reps	Tempo	Sets
Pelvic Shift	L/R 2:00		6-8 each	102	2-4

The Pelvic Shift is performed as in Phase V (Figure 6.18), except the tempo is increased.

The Pelvic Shift exercise runs a cycle beginning with the backswing and concluding with follow-through. The load used should not be so great as to interrupt the normal **RELATIVE TIMING** of the swing motion.

Selecting the correct weight will be influenced by the tempo of execution for the exercise, which is 102; backswing to swings end in a "one thousand and one" count and return to the top of the backswing over two seconds. If you use too much load, then you will have to cheat the timing and disruption of relative timing is inevitable!

It is imperative that you are able to complete 6-8 repetitions with perfect form before feeling as though movement quality is impeded by fatigue. Occasionally, increase the load slightly. If you can go up one plate without disrupting the tempo or relative timing of the exercise, then stick with the new weight.

You will most likely find that a one-plate jump is too great. To remedy this problem use Plate Mates. These are special magnetic micro-plates that stick on the cable column plates, dumbbells or Olympic weights to allow smaller loading increments (see Resources section).

Perform two sets of the Pelvic Shift exercise and progress to four sets. Each set is concluded after performing the exercise to the left and to the right 6-8 times and should be separated by a two minute rest period.

GLOSSARY

Relative Timing: The golf swing can be broken down into five positions according to Bob Cisco: set-up, take away, top of swing, impact and swing's end.[4] The time at which these positions are executed in the swing relative to each other and the overall swing time is "relative timing." Most pro golfers can be identified by the unique relative timing of their swing. Greg Norman, for example, is said to have incredibly consistent relative timing.[5]

Exercise	Rest	Intensity	Reps	Tempo	Sets
Medicine Ball Oblique Toss	L/R 2:00	-2 reps	6-8 each	Fast	2

The Medicine Ball Oblique Toss serves to improve torso rotation power and weight shifting ability. These two movements provide the greatest source of force generation in the golf swing, serving to accelerate the arms and golf club.

The exercise can be performed with the Total Gym Rebounder or against a solid wall with an air-filled medicine ball. The recommended ball weight is 2-8 lbs.

This exercise is performed with very similar relative timing as the Pelvic Shift exercise, except now you also include the arms as an expression of the legs and trunk. Remember, the arms should never be the primary power generator in the golf swing, or any throwing motion, for that matter.

Figure 7.6
Medicine Ball Oblique Toss

- Stand perpendicular to the rebounder or wall.

- Take a stable stance with the ball of your choice in your hands.

- Coil the core and shift your weight so that 70% is on the outside leg and 30% on the inside leg (Figure 7.6A).

- The transition from the coil position to tossing the ball should be fluid; do not stop between the two actions.

- As the body changes direction from coil to toss, draw the navel inward slightly to activate the deep abdominal stabilizer muscles.

- Shift your weight as the ball is tossed Figure 7.6B).

- The arms should be timed such that there is a slight lag, allowing the coil to unwind, driving the arms to a powerful climax as the ball is released.

POWER TRAINING

If the exercise is timed correctly, it will seem effortless and the ball will travel unexpectedly fast. Whether you are using the Total Gym Rebounder or a wall, stand prepared to catch the ball on its return.

The ball should be tossed such that its trajectory on the return places it about 18"-24" out in front of you so that catching the ball pulls you into the coil position. This will effectively prepare you for the next repetition. If you catch the ball too close to your body, the flight of the ball will be interrupted, killing the kinetic energy and reducing the power potential of your next toss.

Becoming proficient at coiling, or storing energy in the muscles, tendons and ligaments as you change directions or phases of the exercise is an important skill to master. As you become more proficient at this process of coiling and releasing, you can rest assured that *your drives will get longer!*

Perform 6-8 repetitions on each side with a fast but rhythmic and controlled tempo. After each set, rest for two minutes. Perform 2-3 sets.

CHAPTER SEVEN

Exercise	Rest	Intensity	Reps	Tempo	Sets
Medicine Ball Lateral Shoulder Rotations	L/R +1:00	-2 reps	6-8 each	Fast	2-4

Medicine Ball Lateral Shoulder Rotations are performed from the neutral position and varying degrees of shoulder abduction up to 80-90°. To keep the exercises as effective as possible, focus your attention on rotation of the upper arm.

Figure 7.7
Medicine Ball Lateral Shoulder Rotations

- Holding a medicine ball, keep your elbow bent to 90°.

- Think of your upper arm as a hinge opening and closing as you rotate your arm. If you are performing the exercise correctly, your upper arm will not wander around, it will just change its relationship to the torso as it progresses from your side up to 80-90° abduction with successive tosses.

- When your arm fatigues to the point that you feel you could do two more tosses before your speed and/or coordination drop, stop and begin a set with the opposite arm. After completing each arm, rest one minute before starting your next set.

If you are using the correct size medicine ball and throwing it with good form, you will reach the end of your set by the time you have performed ten repetitions. It is very important to realize that more is not better with this exercise. The goal is to train the muscle fibers responsible for high-speed contractions – the fibers that are used to stabilize your shoulder as it accelerates the club during the golf swing. If you train into fatigue, you will train yourself to be slower and hit shorter drives!

You will know it is time to use a heavier ball when you are able to perform more than ten reps on each side at high speeds with perfect form and still feel like you have two more good reps left. As with the Medial Shoulder Rotations, increase the weight in small increments so that your speed does not decrease too rapidly.

POWER TRAINING

DAILY PLANNING WORKSHEET

TRAINING PHASE: Phase VI – Power Training for Golf
OBJECTIVE: Increase golf power
DATES: 3-4 weeks

Golf Power – Phase VI

Exercise	Rest	Intensity	Reps	Tempo	Sets
Stretch & Warm Up Set First					
Program A					
Multi-Directional Lunge#	⇓+2:00		2-3 each	101	2-4
Medicine Ball Power Swing	⇓⇨⇧	-2 reps	6-8	101	2-4
Medicine Ball Sit-Up Pass on Swiss Ball	2:00		10-15	101	2-3
Medicine Ball Medial Shoulder Rotations	L/R+1:00	-2 reps	8-10 each	Fast	2-4
Program B					
Medicine Ball Bouncing Wood Chop	⇓+2:00		6-8 each	Fast	2-4
Medicine Ball Reverse Wood Chop	⇓⇨⇧	-1 rep	6-8 each	102	2-3
Pelvic Shift	L/R 2:00		6-8 each	102	2-4
Medicine Oblique Toss	L/R 2:00	-2 reps	6-8 each	Fast	2
Medicine Ball Lateral Shoulder Rotations	L/R+1:00	-2 reps	8-10 each	Fast	2-4

#: Those with knee problems or degenerative knee joints should stick to a slow tempo (303) through this exercise. It will not hinder your golf performance to do this exercise at slow speeds.

CHAPTER SEVEN

GOLF POWER – PHASE VII

Congratulations! You have made it to your final stage of development.

Phase VII is unique because you will not only capitalize on your power potential, but you will also begin restoring static and dynamic stability. The Supine Lateral Ball Roll, the C.R.A.C. Press and Lower Abdominal #2-B Standing will again be performed at this stage. After four weeks of power training, it is normal to begin losing some of your stability skills as well as developing some golf specific muscle imbalances.

To keep your body fine-tuned, balanced and powerful, explosive training will be combined with stabilization and strength exercises.

You should find Phase VII very stimulating. The power exercises are exhilarating and fun to perform. If you ever wanted to compete in a long drive competition, schedule it in the third or fourth week of this phase!

Phase VII Equipment Needs

- Medicine balls ranging from 2-12 lbs. (Some that don't bounce and one that does if you don't have a Total Gym Rebounder.)
- Total Gym Rebounder* (if you do not have an air filled medicine ball)
- Tornado Balls*
- Dumbbells from 5-35 lbs.
- Full-sized mirror with vertical line down center.

*See Resources section on page 219.

POWER TRAINING

PHASE VII EXERCISES

Exercise	Rest	Intensity	Reps	Tempo	Sets
Walking Lunge with Twist	⇩		6-8 each	Moderate	2-3

The Walking Lunge with Twist (LWT) is a great overall conditioning exercise. For the golfer, the LWT serves to further integrate the leg/torso/arm pattern of force and power generation. The exercise also serves to add to the golfer's motor vocabulary while providing a useful training stimulus. Women often like the exercise because it firms the butt!

- Stand with a 2-6 lb. medicine ball positioned by your right side.

- Step into the lunge with your left leg and simultaneously swing your arms in an arc over your head. The movement of your arms should be timed such that it concludes at the same time as the lunge movement with your legs.

- Once you are at the bottom of the lunge with your left leg forward, your torso rotated to the left and the medicine ball just lateral to the left pocket, the motion should continue into the next lunge/twist cycle. Since you are walking forward as you perform the LWT, your momentum will carry you into the next lunge. As you come up from the bottom position, your arms should swing toward the right (arcing overhead) as your right leg steps forward.

Figure 7.8
Walking Lunge with Twist

Your lunge step should be large enough that your front shin does not go forward of vertical. It is common for people to develop pain behind the kneecap if they don't follow this instruction, as allowing the knee to bend excessively under load may place excess strain on the kneecap and its tendons.

As indicated in the Tempo column, this exercise should be performed at a moderate speed. For some athletes the exercise is progressed to the explosive stage. This is not at all necessary for the golfer, particularly if you are over 40 years of age.

Try performing your first set with just your body weight as a warm-up. After your warm-up set, complete one working set with a small medicine ball and see how your body responds over the next 48 hours. This exercise can produce post-exercise soreness, which is also referred to as delayed onset muscle soreness. It has been found that athletes tend to have less delayed onset muscle soreness when consuming anti-oxidants, particularly Vitamin C. You can either take a dose before training or use it for future workouts if you experience soreness.

Count each step as a rep and perform between 12 and 16 LWTs (6-8 on each leg) successively. After completing one set, progress directly to the Standing Tornado Ball Horizontal Chop exercise.

POWER TRAINING

Exercise	Rest	Intensity	Reps	Tempo	Sets
Standing Tornado Ball Horizontal Chop	⇩		<10 secs	XOX	2-4

Now we're getting into the fun stuff! The Standing Tornado Ball Horizontal Chop is the high speed progression of the Pelvic Shift exercise performed in Phase VI.

If this is your first time performing power exercises, use a 1 kg. Twister Ball. The maximum size Tornado Ball for anyone wanting to improve golf performance with this exercise is 3 kg.

- Find a cement wall or a tennis backboard.

- Take a stable stance with your back close to the wall and elbows tucked in at your sides. Make sure you have at least 18" of rope between your hands and the ball. You may wrap the rope around your wrist and back through your hand for extra grip strength. You can also wear weight lifting gloves that cover the wrist for protection.

- From a stable athletic stance, begin to swing the ball left and right bouncing it off the wall. You will have to use enough force to cause the ball to bounce hard off the wall. If you don't hit the wall hard enough, there will not be enough centrifugal force to keep the rope taut and the ball will flutter, making it almost impossible to maintain a rhythm.

Start out by performing a set at a slow pace with the least effort needed to maintain a rhythm. As you warm-up, you can begin to speed the ball up, progressing to the fastest pace you can maintain with good form as shown by the XOX in the Tempo column. Your sets should last no more than ten seconds. This is a power exercise and training longer than ten seconds will fatigue you too much and train you to move slowly; this will not improve your drive.

Figure 7.9
Standing Tornado Ball Horizontal Chop

Once you have the feel of the exercise, it is fun to have contests with friends or training partners to see who can make the ball hit the wall the most times in ten seconds. The person running the stopwatch should start the watch at the instant the first swing makes contact with the wall and should stop it at exactly ten seconds. This is a great power development game that many athletes absolutely love, regardless of which sport they play.

CHAPTER SEVEN

Exercise	Rest	Intensity	Reps	Tempo	Sets
Medicine Ball Soccer Toss	⇩⇨⇧		8-12	X	2-3

The Medicine Ball Soccer Toss continues your core conditioning and power development. The exercise can be performed against a Total Gym Rebounder, against a wall with an air filled medicine ball or with a partner of comparable strength.

- Hold a 2-8 lb. medicine ball directly over your head, with your feet no wider than hip width apart.

- Toss the ball at the rebounder. Activate the abdominals as you toss the ball so that you do not overuse your arms.

- When you toss the ball, it should bounce back slightly above head level. This will produce an extension (backward bending) force on the trunk which will have to be controlled by the abdominal muscles. This again is econcentric training and serves to stretch the spring for an explosive return. This is exactly the kind of strength you need at the top of your backswing.

Figure 7.10
Medicine Ball Soccer Toss

You should be able to perform 8-12 reps explosively (see X in Tempo column). You will notice there is only one X in the tempo column. That is because the exercise consists primarily of the explosive toss. The catch phase of the exercise is not recorded in the tempo column because it is not meant to be performed at any specific tempo, just in a controlled manner without stopping the ball. If you stop the ball when you catch it, you will not be developing the type of strength needed to increase your drive!

After performing one set, rest for two minutes and start another mini-circuit beginning with the Walking Lunge with Twist.

POWER TRAINING

Exercise	Rest	Intensity	Reps	Tempo	Sets
Rotator Cuff Tornado Ball Cross Body Pattern	L/R+1:00		<10 sec	XOX	2-3

The Tornado or Twister ball can be effectively utilized to condition the rotator cuff musculature of the shoulder. This is done by combining the movements of abduction and external rotation with adduction and internal rotation. It is important to progress rotator cuff conditioning exercises to this level of integration because these are common movement patterns in most work and sports environments. For the golfer, these patterns are similar to the backswing and downswing patterns of motion with regard to rotator cuff activation.

- The exercise should be performed with a Twister Ball weighing between 0.5-1 kg.

- Secure your hold on the rope with about 18" of rope between the ball and your hand.

- Move your arm downward and across your body.

- When the ball bounces off the wall, you should immediately draw the ball upward and back at a 45° angle, creating the complete cross-body pattern.

After performing an easy set or two on each arm as a warm-up, progress to the work sets. Your sets should last no longer than ten seconds per arm. The tempo is XOX which indicates an explosive movement in both directions. The movement should never be performed so fast that you are not in complete control! After performing a set on each arm, rest for one minute before beginning the next set. A total of 2-3 working sets should be performed after your warm-up.

Figure 7.11
Rotator Cuff Tornado Ball Cross Body Pattern

CHAPTER SEVEN

Exercise	Rest	Intensity	Reps	Tempo	Sets
Tornado Ball Pelvic Shift	2:00		6-8 each	XOX	2-4

The Tornado Ball Pelvic Shift exercise is a progression from the standard Pelvic Shift exercise performed in Phase VI. The Tornado Ball allows much faster weight transfer. Faster weight transfer and trunk movement provide the perfect environment to express the strength developed with the Pelvic Shift in Phase VI as power.

- Take a stable stance with your back close to a cement wall or a tennis backboard.

- Keep your elbows tucked in at your sides at all times so that you do not hit an elbow against the wall.

- Hold the rope just in front of your pelvis right above your pubic bone. Keep your hands in this position throughout the exercise.

- Shift your weight laterally and rotate your pelvis from side-to-side to swing the ball against the wall. As you perform the motion, visualize the exact hip and leg motion used during the golf swing. This will help sequence your muscles for the highest level of strength and power transfer to golf. Emphasize the pelvic motion and de-emphasize the movement of the trunk.

Figure 7.12
Tornado Ball Pelvic Shift

Initially, start slowly to get a rhythm going. For the first set, keep the effort below 60% to warm up. For the next set, start out at a comfortable pace and speed up with every repetition until you are going at maximum pace.

Having completed your set, rest for two minutes. You should complete 2-4 sets before progressing to the Medicine Ball Power Swing. The Medicine Ball Power Swing should feel very fluid after having performed the Tornado Ball Pelvic Shift. From a neurological perspective, the progression of these exercises serves as an integrative stimulus for your nervous system because the Medicine Ball Power Swing builds upon the Tornado Ball Pelvic Shift.

Exercise	Rest	Intensity	Reps	Tempo	Sets
Medicine Ball Power Swing	2:00		6-8 each	XOX	2-3

The Medicine Ball Power Swing exercise is performed as instructed in Phase VI, Program A. The only change

POWER TRAINING

you should take note of is that the tempo has changed from 101 (or fast) to XOX, which is explosive. You should warm-up by performing the Medicine Ball Power Swing exercise for one set at 60% effort followed by a second set which progresses from 60% on the first repetition to 90% of your working or actual training effort by the eighth repetition.

A very important facet of this exercise is maintaining your swing axis. If you are explosive without concomitant control, then you will only train yourself to hit hooks and/or slices! Like golf, these exercises require graceful expression of power.

Exercise	Rest	Intensity	Reps	Tempo	Sets
Single Arm C.R.A.C. Press	1:00		6-8 each	303	1-3

The Single Arm C.R.A.C. Press is performed exactly as it was in Phase V, Program A. Complete the exercise on both arms then rest one minute.

Exercise	Rest	Intensity	Reps	Tempo	Sets
Supine Lateral Ball Roll	1:00		6-8 each	1 sec hold	2-3

The Supine Lateral Ball Roll is performed exactly as outlined in Phase III. This dynamic stability exercise is included in the power phase because any time you maximize one quality in training you begin to lose another quality. For the golfer, losing dynamic stability at the expense of gaining power is dangerous. Everyone loves to hit the ball long, but who wants to claim the record for hitting a ball farthest into the trees!

The Supine Lateral Ball Roll will serve as an excellent antidote for your new explosive capabilities, allowing you to continue your power development while maintaining dynamic stability. Having power without dynamic stability would be like having a 500 horsepower speedboat without a rudder!

Exercise	Rest	Intensity	Reps	Tempo	Sets
Lower Abdominal #2-B Standing Supported	1:00		2-20 each	S-M-F	2-3

The Lower Abdominal #2-B Standing Unsupported exercise is performed exactly as described in Phase V, Program B, with a one minute rest between sets.

CHAPTER SEVEN

DAILY PLANNING WORKSHEET

TRAINING PHASE: Phase VII – Power Training for Golf
OBJECTIVE: Increase golf power
DATES: 3-4 weeks

Golf Power – Phase VII

Exercise	Rest	Intensity	Reps	Tempo	Sets
Stretch & Warm Up Set First					
Program A					
Walking Lunge with Twist[π]	⇩+2:00		6-8 each	Moderate	2-3
Standing Tornado Ball Horizontal Chop	⇩		10 sec	XOX	2-4
Medicine Ball Soccer Toss	⇩⇨⇧		8-12	X	2-3
Rotator Cuff Tornado Ball Cross Body Pattern	L/R+1:00		10 sec	XOX	2-3
Program B					
Tornado Ball Pelvic Shift	2:00		6-8 each	XOX	2-4
Medicine Ball Power Swing	2:00		6-8 each	XOX	2-3
Single Arm C.R.A.C. Press	1:00		6-8 each	303	1-3
Supine Lateral Ball Roll	1:00		6-8 each	1 sec hold	2-3
Lower Abdominal #2-B Standing Unsupported	1:00		12-20 each	S-M-F	2-3

π: Those with knee problems or degenerative knee joints should stick to a slow tempo (303) through this exercise. It will not hinder your golf performance to do this exercise at slow speeds.

CHAPTER 8

NOW THAT YOU'RE HOOKED!

Congratulations! You have just completed seven months of the world's most comprehensive golf conditioning program. You may be wondering what to do from here. Don't worry, your training programs for the rest of the year are outlined in this chapter. No doubt you have seen many positive changes in your game since beginning the program. You have most likely noticed that your posture has improved, your balance is better, coordination and strength levels are at, or exceed, all time highs and you are hitting the ball farther than ever.

You don't want to stop now. As alluded to in Phase VII, when you become more specific in any conditioning program, you always lose a degree of another quality or qualities. At the end of Phase VII, your power levels will be very high, yet you may be showing signs of decreased static and dynamic stability. Another factor by the end of Phase VII is strength loss.

Focusing on power development doesn't allow for strength development. In many cases athletes lose strength as power goes up. This is often the case with basketball players and volleyball players who get very good at jumping as their power increases but eventually, as their strength levels fall, they can no longer develop the power to jump as high and they must return to a cycle of strength training.

Your goal now will be to maintain as much power as possible while cycling back to stabilization and strength training. Cycling back is important because you have just completed one full cycle of development. From here, your goal is to become more skilled and stronger in each phase than you were the first time.

> *Your goal now will be to maintain as much power as possible while cycling back to stabilization and strength training.... From here, your goal is to become more skilled and stronger in each phase than you were the first time.*

CHAPTER EIGHT

As you are aware, an author must assume a tremendous amount about his readers. If you were being conditioned at the C.H.E.K Institute, you would be going through re-evaluations every four weeks and getting scientifically engineered programs for your specific needs. In absence of such specificity, you have been given program proposals that should serve as a technical guideline.

Before beginning Phase VIII, perform all the length/tension tests again to determine which stretches you need to continue performing and which ones are no longer necessary at this point. Time is valuable and you don't want to spend it stretching muscles that are functioning perfectly well. Instead, you want to focus on the areas of greatest concern.

The Areas of Greatest Concern

By now you will have determined which areas you need the most help with in your golf game (and your body!).

- If you are driving the ball record distances compared to before starting this program, but are not putting as well as you were in the past three or four months, then you will need more static and dynamic stability training.

- If your putting is great, but you are having a hard time chipping or pitching the ball, then focus on dynamic stability.

- If you feel you have improved your drive, but relative to the rest of your game it is still your biggest weakness, then focus on another cycle of strength exercises, maybe even two cycles, before returning to Phase VI and Phase VII again.

You will be pleasantly surprised at how much easier the exercises are the second time around. Your body and mind will be pre-framed for success coming into your next phase of conditioning. You will know exactly what to do in the gym and what your goals are.

If you feel your physical condition has notably improved, yet your score is not improving at the rate you feel it should, then it may be the right time to ask for help from a successful golf pro! The primary purpose of your Whole In One Golf Conditioning program has been to get your body in balance and conditioned to allow it to react favorably to your mind. If your body is now performing well and you are feeling well, but not playing as well as you feel you should be, it is the golf pro's job to reprogram your new body with good golf technique and strategy.

It is also good to help your mind! One of the best ways to do that is to read the best books you can find on the strategic and mental aspects of the golf game to accent your physical attributes. An excellent book on this topic is *The Ultimate Game of Golf* by Bob Cisco.[1]

NOW THAT YOU ARE HOOKED

You may resort to any one of the previous seven phases of conditioning. If you feel that any particular phase impacted your game more positively than the others, return to that phase for three or four weeks and then progress through the successive phases again. You will only be better off the second time around.

The Remainder of Your Conditioning Year

The following five conditioning cycles may be used for up to four weeks each. Each golf skill area is represented with a suggested conditioning schedule to finish the year (Table 8.3). These conditioning programs are designed to foster progressive development in your skill area of concern and your golfing body.

It is very normal for clients using the Whole In One exercise program to have resolution of low back pain, knee pain, shoulder pain, headaches and many other seemingly unrelated symptoms and ailments. This is because Whole In One Golf Conditioning is built entirely upon the principles of corrective exercise and kinesiology. This approach to conditioning any athlete or patient is to first re-establish function and then performance. Resolution of the above complaints is secondary to restoration of movement skill and function.

Staying Sharp

Having made it this far in your program is evidence of your commitment to excellence! You are no doubt very aware of which areas of your game need improvement. As mentioned in the beginning of this book, the five factors that control ball flight are directly related to four key physical factors (Table 8.1). Common swing faults that result from deficits in one or more of these physical factors are shown in Figure 8.1.

As you can see in Table 8.1, certain ball flight factors are highly correlated to particular physical factors. For example club head speed is highly correlated to strength and power. Similar correlations can be made between the different skills required for golf success and the physical attributes most necessary to maintain particular golf skills (Table 8.2).

By assessing which golf skill needs the most improvement upon completion of Phase VII, you can more effectively focus your conditioning efforts in Phase VIII. For example, if putting is your main area of concern, Phase VIII should primarily focus on static and dynamic stability (see Table 8.2). If distance is keeping you from achieving your scoring potential and the rest of your game is intact, you need to reassess your flexibility, maintain dynamic stability, cycle through another eight weeks of strength training and then another four to eight weeks of power training.

To guide and direct you through the remainder of the year and beyond, follow the recommendations in Table 8.3.

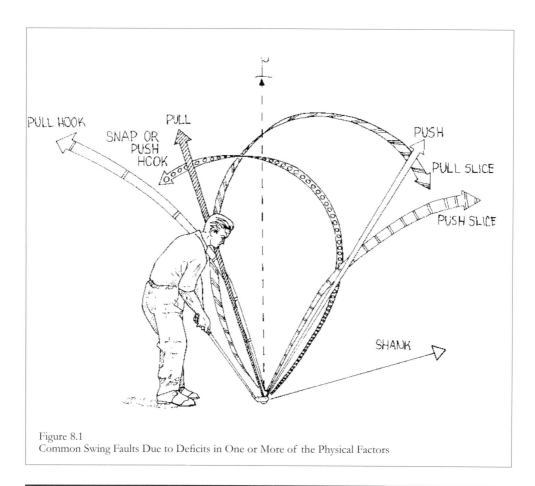

Figure 8.1
Common Swing Faults Due to Deficits in One or More of the Physical Factors

Ball Flight Factors	Physical Factors
Clubface Alignment Swing Path Angle of Attack Hitting Sweet Spot	Flexibility Static & Dynamic Stability
Speed	Strength Power

Table 8.1

Recommendations for the Serious Amateur or Professional Golfer

Golf Skill	Physical Attribute
Putting	Static and Dynamic Stability
Chipping	Flexibility/Static and Dynamic Stability/Strength
Pitching	Flexibility/Static and Dynamic Stability/Strength
Bunker	Flexibility/Static and Dynamic Stability/Strength
Driving	Flexibility/Static and Dynamic Stability/Strength/Power

Table 8.2
Golf Skill/Physical Attribute Correlations

Skill Focus	Phase VIII	Phase IX	Phase X	Phase XI	Phase XII
Putting	Repeat Phase II	Repeat Phase III	Repeat Phase IV	Repeat Phase V	Repeat Phase VI
Chipping/ Pitching/ Bunker	Repeat Phase III	Repeat Phase IV	Repeat Phase V	Repeat Phase VI	Repeat Phase VII
Driving	Repeat Phase IV	Repeat Phase V	Repeat Phase VI	Repeat Phase VII	Repeat Phase II-VII

Table 8.3
Programming to Match Skill Requirements with Physical Attributes for Successful Golf Conditioning

CHAPTER EIGHT

A devoted student of golf and golf conditioning will make great strides toward his or her lowest handicap ever by combining the conditioning programs in this book with the services of a good teaching pro.

To teach the golfer with little or no experience with scientific conditioning for the sport is very challenging, as you can imagine. I have done my best to describe and periodize the exercise programs in this book with the knowledge that the information must be useful to golfers of all levels of conditioning and golf skills. To truly reach your golf potential as an elite or professional golfer, I suggest consulting a C.H.E.K Practitioner or C.H.E.K CHEK Golf Biomechanic (visit the online database at www.CHEKconnect.com).

Through testing and professional guidance, you can achieve far more than you could ever achieve through reading a book. For example, can you imagine ever reaching an elite level as a golfer without professional instruction and just by reading a book?

When I began writing this book, I did so because I was amazed at the lack of information on conditioning available to golfers. I was compelled to do my best to introduce the golf world to the type of conditioning I have used for over seventeen years to rehabilitate and enhance performance in golfers and other professional athletes. With successful application of scientific conditioning principles, today's golfers may, for the first time in over thirty years, approach lowering their golf scores.

As echoed in the beginning of the book,
THE GOLFER PLAYS THE GAME, NOT THE CLUBS!

RESOURCES

C.H.E.K Institute-trained Professionals

C.H.E.K Institute-trained Professionals are fitness and health professionals who have completed one or more levels of the C.H.E.K Institute's Advanced Training Programs. Depending on your needs, you may find it helpful to consult with one of these specialists during your Whole in One training program.

- CHEK Golf Biomechanics specialize in conditioning programs for golfers
- CHEK Exercise Coaches and C.H.E.K Practitioners (Corrective High-performance Exercise Kinesiology Practitioners) work in the areas of corrective exercise, high-performance conditioning and more advanced rehabilitation and post-rehabilitation exercise, depending on their level of training and prior education.
- CHEK Holistic Lifestyle Coaches (HLCs) can help with stress management and healthy lifestyle principles including nutrition, hydration, work-rest balance and more.

To find a C.H.E.K Institute-trained Professional in your area, please check the online database at www.CHEKconnect.com. Alternatively, you can contact the C.H.E.K Institute at info@chekinstitute.com or call 1.800.552.8789.

C.H.E.K Institute International Distributors

The C.H.E.K Institute has international distributors in many regions of the world who carry books, DVDs and equipment recommended in this book. For an up-to-date list please visit www.CHEKdistributors.com.

Equipment, Books and DVDs

Many of the items listed below are available from the C.H.E.K Institute online store at www.chekinstitute.com – then click the Shop tab.

Chapter 2 - Flexibility Assessment

Blood Pressure Cuff – you will want a blood pressure cuff with a manual gauge rather than a digital one. These can be harder to find nowadays, although your local pharmacy may still stock them. You can alternatively use a biofeedback unit specifically designed for lower abdominal assessment and conditioning. Both are available from the C.H.E.K Institute's online store in the *Golf* department.

Chapter 3 - Stretching

Wooden Dowel Rod – these are actually closet rods and available from most home supply stores. Choose one that is 1 3/8" (3.5cm) in diameter and have it cut to 6ft (2m) long. Stand down the rough ends to avoid splinters.

RESOURCES

Foam Roller – you want to find one that is 3" or 4" in diameter. The more common size is 6" in diameter, but this is too large for the mobilization exercises. Available from the C.H.E.K Institute's online store in the *Golf* department.

Swiss Ball – also know as a stability ball or physio ball. Make sure you select a burst-resistance ball, with a burst resistant rating that is more than your body weight e.g. if you weigh 200 lbs, the ball must be rated at least 200 lbs, and preferably 250lbs or more. Many cheap balls are not burst resistance and will pop like a balloon if they get punctured, which is not a good idea if you happen to be sitting on it at the time! Available from the C.H.E.K Institute's online store in the *Golf* department.

Athletic Tape – any strong, sticky tape will work, but it must not stretch. If you are allergic to the adhesive, some brands include a backing tape to apply to the skin and then the athletic tape goes on top. The EnduraTape brand is available from the C.H.E.K Institute's online store in the *Golf* department.

Chapter 5 – Functional Exercise

Wooden Dowel Rod – see above
Swiss Ball – see above
Blood Pressure Cuff – see above

Gertie Ball – a small inflatable ball about 1 ft to 1½ ft (30-45cm) in diameter. Available from the C.H.E.K Institute's online store in the *Golf* department.

PhysioToner – a loop of elastic resistance tubing with foam grips and secured in the middle to look like a figure 8. Available in different lengths and resistances from the C.H.E.K Institute's online store in the *Golf* department.

Power Web – a circular rubber web used for conditioning the fingers and hand. Available in different resistances from the C.H.E.K Institute's online store in the *Golf* department.

Fitter Wobble Board – a circular wooden board with an adjustable height setting. Start with the 20" diameter board; the 16" board is much harder. If the wobble board is too challenging to begin, try the Fitter Rocker Board instead. Available in different resistances from the C.H.E.K Institute's online store in the *Golf* department.

Pro Fitter - Available from the C.H.E.K Institute's online store in the *Golf* department.

Useful DVDs for this chapter containing many of the exercises demonstrated are available from the C.H.E.K Institute's online store in the *DVD* department:

- Paul Chek's Gym Instructor Series Volume 1 Core Conditioning, Part 1 Abdominal Training
- Paul Chek's Gym Instructor Series Volume 1 Core Conditioning, Part 2 Back and Ball Training

- Paul Chek's Gym Instructor Series Volume 3 Rows, Pulls, Chins and the Dead Lift
- Scientific Back Training 5-DVD set
- Swiss Ball Exercises for Better Abs, Buns and Backs
- Swiss Ball Exercises for Athletes
- High Performance Core Conditioning 2-DVD set

Chapter 6 – Strength Training

Swiss Ball – see above
Blood Pressure Cuff – see above
Wooden Dowel Rod – see above

Adjustable Cable Machine – there are several models available on the market. The cable position needs to be able to adjust from high to low and all positions in between. The FreeMotion Cable Column (www.freemotionfitness.com in their commerical-studio line of equipment) is great for home gyms as it can be easily moved around on its wheels. An alternative is a Total Gym or Gravity machine, both available from efi sports medicine (www.efisportsmedicine.com). See also the Resources section of the C.H.E.K Institute website at www.chekinstitute.com for other options.

Box Step – any platform with adjustable risers will work. Most sporting goods stores have steps that are used for group exercise step classes, and you may want to purchase some additional risers. See also the Resources section of the C.H.E.K Institute website at www.chekinstitute.com for other options.

Adjustable Dead Lift Blocks – you can see these in use in Figure 6.9. Find suppliers listed in the Resources section of the C.H.E.K Institute website at www.chekinstitute.com.

Olympic Bar and Weights – there are many manufacturers of these. At the C.H.E.K Institute we use Eleiko equipment (www.eleikosport.se); the highest quality weight-lifting equipment in the world. You will also be able to find bars and weights at your local sporting goods store, or check out the Resources section of the C.H.E.K Institute website at www.chekinstitute.com.

Useful DVDs for this chapter containing many of the exercises demonstrated are available from the C.H.E.K Institute's online store in the *DVD* department:

- Paul Chek's Gym Instructor Series Volume 1 Core Conditioning, Part 1 Abdominal Training
- Paul Chek's Gym Instructor Series Volume 2 Pushing and Pressing Exercises
- Paul Chek's Gym Instructor Series Volume 3 Rows, Pulls, Chins and the Dead Lift
- Scientific Back Training 5-DVD set
- Equal, But Not the Same: Considerations for Training Females 5-DVD set
- High Performance Core Conditioning 2-DVD set
- Strong 'N' Stable: Swiss Ball Weight Training 3-DVD set

RESOURCES

Chapter 7 – Power Training

Wooden Dowel Rod – see above
Swiss Ball – see above
Blood Pressure Cuff – see above

Olympic Bar and Weights – see above

PlateMates – small, magnetic weights that you can attach to dumbbells or weight stacks and increase the resistance a fraction, rather than having to struggle with the next size dumbbell or weight plate. www.theplatemate.com or find suppliers listed in the Resources section of the C.H.E.K Institute website at www.chekinstitute.com.

Medicine Balls – these come in various different weights and sizes, so you will want to choose the correct ball for each exercise. If you are throwing and catching with one hand, you will want a smaller Gripper size ball. For two handed exercises, you can use a larger basketball-size ball. There are also medicine balls that bounce well – best used when you are throwing against a wall or floor and want the ball to come back to you – and dead medicine balls that do not bounce, which are best used with the rebounder. Dead balls are available from the C.H.E.K Institute's online store in the *Golf* department. Find suppliers for other medicine balls listed in the Resources section of the C.H.E.K Institute website at www.chekinstitute.com.

Total Gym Plyometric Rebounder – Available from the C.H.E.K Institute's online store in the *Golf* department.

Tornado and Twister Balls – weighted balls with a rope securely attached to them. Available from the C.H.E.K Institute's online store in the *Golf* department.

Also Available from the C.H.E.K Institute

Golf Card Set

120 cards with all the assessments stretches and exercises featured in this book. Easy to carry with you to the golf course or gym - select the cards you need and attach to the enclosed ring.

A Scientific Approach to Golf Conditioning DVD - (108 min. lecture)

Join Paul Chek as he explains why most golf training programs fail to reduce the risk of injury or improve performance. A technical, yet fascinating talk!

Visit www.chekinstitute.com to order or find your local distributor at www.chekdistributors.com

REFERENCES

Chapter 1

1. Watkins, R.G., Uppal, G.S., Perry, J., Pink, M. & Dinsay, J.M. (1996). Dynamic Electromyographic Analysis of Trunk Musculature in Professional Golfers. *The American Journal of Sports Medicine, 24,* (4).
2. McCarroll, J.R. (1996). The Frequency of Golf Injuries. *Clinics in Sports Medicine,* 15(1).
3. Fore! Athletes who Golf are at Risk for Back Pain. (November 1996). *The Joint Letter, 2*(10).
4. Williams, M. (Ed.). (1995). *The Royal and Ancient Golfer's Handbook 1995.* London, UK: Macmillan.
5. Johnson, S. & Seanor, D. (2009). The USA Today Golfers Encyclopedia. New York, NY: Skyhorse Publishing.
6. Rotella, R. (1997). *The Golf of Your Dreams.* Simon and Schuster.
7. Schmidt, R. H. (1991). *Motor Learning and Performance.* Champaign, IL: Human Kinetics. 92-109.
8. Shapiro, D.C. (1978). *The Learning of Generalized Motor Programs.* Ph.D. Dissertation, University of Southern California.
9. Steindler, A. (1955). *Kinesiology of the Human Body.* Springfield, IL: Charles C. Thomas.
10. Goldthwait, J.E., Brown, L.T., Swaim, L.T. & Kuhns, J.G. (1952). *Essentials Of Body Mechanics In Health and Disease* (5th ed.). Philadelphia, PA: J.B. Lippincott Company.
11. Lowman, C. L. & Young, C. H. (1960). *Postural Fitness – Significance and Variances.* Lea & Febiger.
12. Hogan, B. (1957 & 1985). *Ben Hogan's Five Lessons – The Modern Fundamentals of Golf.* Trumbull, CT: NYT Special Services Inc.
13. Kendall, H.O., Kendall, F.P. & Boynton, D.A. (1952). *Posture and Pain.* Malabar, FL: Williams and Wilkins.
14. Chek, P. & Curl, D. (1994). Posture and Craniofacial Pain. In: *Chiropractic Approach To Head Pain,* Curl, D.D. (ed.). Baltimore, MA: Williams and Wilkins. 121-162.
15. Feldenkrais, M. (1949). *Body and Mature Behavior.* Madison, CT: International Universities Press.
16. Hogan, B. (1957 & 1985). *Ben Hogan's Five Lessons – The Modern Fundamentals of Golf.* Trumbull, CT: NYT Special Services Inc.
17. Chek, P. (1993). *Scientific Back Training Correspondence Course.* Encinitas, CA: C.H.E.K Institute.
18. Cisco, B. (1993). *The Ultimate Game of Golf.* Glendale, CA: Griffin Publishing.
19. McCarroll, J.R. (1996). The Frequency of Golf Injuries. *Clinics in Sports Medicine,* 15(1).
20. McCarroll, J.R. (1996). The Frequency of Golf Injuries. *Clinics in Sports Medicine,* 15(1).
21. Chek, P. & Curl, D. (1994). Posture and Craniofacial Pain. In: *Chiropractic Approach To Head Pain,* Curl, D.D. (ed.). Baltimore, MA: Williams and Wilkins. 121-162.
22. Seamann, D. (1978). C1 Subluxations, Short Leg and Pelvic Distortions. *Upper Cervical Monograph, 2*(5).
23. Chek, P. & Curl, D. (1994). Posture and Craniofacial Pain. In: *Chiropractic Approach To Head Pain,* Curl, D.D. (ed.). Baltimore, MA: Williams and Wilkins. 121-162.
24. Gregory, RR. (1984). Updating the Basic Types. *Upper Cervical Monograph, 3,* 8 –12.
25. Chek, P. & Curl, D. (1994). Posture and Craniofacial Pain. In: *Chiropractic Approach To Head Pain,* Curl, D.D. (ed.). Baltimore, MA: Williams and Wilkins. 121-162.
26. Seamann, D. (1978). C1 Subluxations, Short Leg and Pelvic Distortions. *Upper Cervical Monograph, 2*(5).
27. Chek, P. & Curl, D. (1994). Posture and Craniofacial Pain. In: *Chiropractic Approach To Head Pain,* Curl, D.D. (ed.). Baltimore, MA: Williams and Wilkins. 121-162.
28. Feldenkrais, M. (1949). *Body and Mature Behavior.* Madison, CT: International Universities Press. 66-69.

REFERENCES

29. Chek, P. (1993). *Scientific Back Training Correspondence Course*. Encinitas, CA: C.H.E.K Institute.

Fig.1.14 Lemmon, G.J. (1941). *About Golf*. Publisher unknown.

Chapter 2
1. Spring, H. et. al. (1991). *Stretching and Strengthening Exercise*. Thieme Publishers.
2. Chek, P. & Curl, D. (1994). Posture and Craniofacial Pain. In: *Chiropractic Approach To Head Pain*, Curl, D.D. (ed.). Baltimore, MA: Williams and Wilkins. 121-162.
3. Hoppenfeld, S. (1976). *Physical Examination of the Spine and Extremities*. Norwalk, CN: Appleton-Century-Crofts.
4. Jobe, F.W. & Pink, M.M. (1996). Shoulder Pain in Golf. *Clinics in Sports Medicine, 15*(1).

Chapter 3
1. Schmidt, R. H. (1991). *Motor Learning and Performance*. Champaign, IL: Human Kinetics. 92-109.
2. Shapiro, D.C. (1978). *The Learning of Generalized Motor Programs*. Ph.D. Dissertation, University of Southern California.

Chapter 4
1. Cisco, B. (1993). *The Ultimate Game of Golf*. Glendale, CA: Griffin Publishing.

Chapter 5
1. Jull, G.A. & Janda, V. (1987). Muscles and Motor Control in Low Back Pain: Assessment and Management. In: *Physical Therapy of the Low Back*. Twomey, L.T. & Taylor, J.R. (Eds). Churchill Livingstone.
2. Hodges, P.W. & Richardson, C.A. (1997).Contraction of the Abdominal Muscles Associated With Movement of the Lower Limb. *Physical Therapy, 77*(2).
3. Hodges, P.W. & Richardson, C.A. (1997). Feedforward Contraction Of Transverse Abdominus is not Unfluenced by the Direction of Arm Movement. *Experimental Brain Research, 114*, 362-370.
4. Hosea, T.M & Gatt, C.J., Jr. (1996). Back Pain in Golf. *Clinics in Sports Medicine, 15*(1).
5. McCarroll, J.R. (1996). The Frequency of Golf Injuries. *Clinics in Sports Medicine, 15*(1).
6. Hodges, P.W. & Richardson, C.A. (1997).Contraction of the Abdominal Muscles Associated With Movement of the Lower Limb. *Physical Therapy, 77*(2).
7. Hodges, P.W. & Richardson, C.A. (1997). Feedforward Contraction Of Transverse Abdominus is not Influenced by the Direction of Arm Movement. *Experimental Brain Research, 114*, 362-370.
8. Hosea, T.M & Gatt, C.J., Jr. (1996). Back Pain in Golf. *Clinics in Sports Medicine, 15*(1).
9. Vleeming, A., Mooney, V., Dorman, T., Snijders, C. & Stoeckart, R. (1997). *Movement, Stability and Low Back Pain*. Edinburgh, UK: Livingston Churchill.
10. Chek, P. (1998). *Scientific Core Conditioning Correspondence Course*. Encinitas, CA: C.H.E.K Institute.
11. Goldthwait, J.E., Brown, L.T., Swaim, L.T. & Kuhns, J.G. (1952). *Essentials Of Body Mechanics In Health and Disease* (5th ed.). Philadelphia, PA: J.B. Lippincott Company.
12. Hodges, P.W. & Richardson, C.A. (1997).Contraction of the Abdominal Muscles Associated With Movement of the Lower Limb. *Physical Therapy, 77*(2).

13. Hodges, P.W. & Richardson, C.A. (1997). Feedforward Contraction Of Transverse Abdominus is not Influenced by the Direction of Arm Movement. *Experimental Brain Research, 114,* 362-370.
14. Vleeming, A., Mooney, V., Dorman, T., Snijders, C. & Stoeckart, R. (1997). *Movement, Stability and Low Back Pain.* Edinburgh, UK: Livingston Churchill.
15. Chaffin, D.B. &. Andersson, G.B.J. (1991). *Occupational Biomechanics* (2nd ed.). New York: Jahn Wiley and Sons, Inc.
16. Schmidt, R. H. (1991). *Motor Learning and Performance.* Champaign, IL: Human Kinetics. 92-109.

Chapter 6

1. McCarroll, J.R. (1996). The Frequency of Golf Injuries. *Clinics in Sports Medicine, 15*(1).
2. Burdorf, A., Van Der Steenhoven, G.A. & Tromp-Klaren, E.G.M. (1996). A One-Year Prospective Study on Back Pain Among Novice Golfers. *American Journal of Sports Medicine, 24*(5).
3. Watkins, R.G., Uppal, G.S., Perry, J., Pink, M. & Dinsay, J.M. (1996). Dynamic Electromyographic Analysis of Trunk Musculature in Professional Golfers. *The American Journal of Sports Medicine, 24,* (4).
4. McCarroll, J.R. (1996). The Frequency of Golf Injuries. *Clinics in Sports Medicine, 15*(1).
5. Hosea, T.M & Gatt, C.J., Jr. (1996). Back Pain in Golf. *Clinics in Sports Medicine, 15*(1).
6. Athletes, Backs, and Golf. (1996, November). *The Joint Letter, 2*(10), 111.
7. Hosea, T.M & Gatt, C.J., Jr. (1996). Back Pain in Golf. *Clinics in Sports Medicine, 15*(1).
8. Hodges, P.W. & Richardson, C.A. (1997).Contraction of the Abdominal Muscles Associated With Movement of the Lower Limb. *Physical Therapy, 77*(2).
9. Hodges, P.W. & Richardson, C.A. (1997). Feedforward Contraction Of Transverse Abdominus is not Unfluenced by the Direction of Arm Movement. *Experimental Brain Research, 114,* 362-370.
10. Chek, P. (1998, May). Is Your Routine Out Of Order? *Muscle Media,* 136-143.
11. Schmidt, R. H. (1991). *Motor Learning and Performance.* Champaign, IL: Human Kinetics. 92-109.
12. McCarroll, J.R. (1996). The Frequency of Golf Injuries. *Clinics in Sports Medicine, 15*(1).
13. Chek, P. & Curl, D. (1994). Posture and Craniofacial Pain.
In: *Chiropractic Approach To Head Pain,* Curl, D.D. (ed.). Baltimore, MA: Williams and Wilkins. 121-162.
14. Chek, P. (1998). *Advanced Program Design Correspondence Course.* Encinitas: C.H.E.K Institute.
15. McCarroll, J.R. (1996). The Frequency of Golf Injuries. *Clinics in Sports Medicine, 15*(1).
16. Burdorf, A., Van Der Steenhoven, G.A. & Tromp-Klaren, E.G.M. (1996). A One-Year Prospective Study on Back Pain Among Novice Golfers. *American Journal of Sports Medicine, 24*(5).
17. Hosea, T.M & Gatt, C.J., Jr. (1996). Back Pain in Golf. *Clinics in Sports Medicine, 15*(1).
18. Athletes, Backs, and Golf. (1996, November). *The Joint Letter, 2*(10), 111.
19. Cavallo, R.J. & Speer, K.P. (1998). Shoulder Instability and Impingement in Throwing Athletes. *Medicine and Science in Sports and Exercise, 30*(4).
20. Kibler, W.B. (1998). Shoulder Rehabilitation: Principles and Practice. *Medicine and Science in Sports and Exercise, 30*(4).

Chapter 7

1. Kendall, H.O., Kendall, F.P. & Boynton, D.A. (1952). *Posture and Pain.* Malabar, FL: Williams and Wilkins.

REFERENCES

2. Watkins, R.G., Uppal, G.S., Perry, J., Pink, M. & Dinsay, J.M. (1996). Dynamic Electromyographic Analysis of Trunk Musculature in Professional Golfers. *The American Journal of Sports Medicine, 24*, (4).
3. Kao, J.T., Pink, M., Jobe, F.W. & Perry, J. (1995). Electromyographic Analysis of the Scapular Muscles During a Golf Swing. *The American Journal of Sports Medicine. 23*(1).
4. Steindler, A. (1955). *Kinesiology of the Human Body.* Springfield, IL: Charles C. Thomas.
5. Lowman, C. L. & Young, C. H. (1960). *Postural Fitness — Significance and Variances.* Lea & Febiger.

Chapter 8
1. Steindler, A. (1955). *Kinesiology of the Human Body.* Springfield, IL: Charles C. Thomas.

INDEX

A

abdominals 121,133, 135, 155, 193
    lower abdominal #1-3 111-114, 116-118, 174-75, 186, 187, 201, 211-12
    obliques 57,
    oblique toss 200-201, 203
    rectus abdominus 56
Adjustable block 170
agility 136
angle of attack 4, 29-30, 71, 189, 190, 216
ankle warm-up 88
ankylosing spondylitis 42
anti-oxidants 206
Apley scratch test 32
arm flicks 89
arm raise test 40
axis of rotation 14, 15, 96, 127

B

back muscles 25
back pain 2, 35, 38, 91, 131, 135, 142, 144, 170
backswing 14, 15, 21, 28, 33, 35, 39, 41, 43, 71, 87, 104, 140, 209
balance 91, 97, 135, 139, 142-3, 204
ball flight 4, 185, 215, 216
bent over row 125, 127-129
Big Bang exercise 136
blood pressure cuff 36, 98, 110-14, 153
box step-up 162-3, 175
bunker 217

C

cable pulley 177
calves 67
cardiovascular exercise 115
center of gravity 5-6
cervical discs 49
chipping 214, 217
cigarette butt stretch 62
cigarette butt test 38
closed chain exercise 5, 7
clubface alignment 4, 40, 189, 190, 216
club path 96
club speed 189, 190, 215
coil 15, 34, 39, 41, 83, 138, 197, 200-201
core strength 136, 141, 185
C.R.A.C. (Core Recruitment Antagonist Co-Contraction) Press 180
cross body external shoulder rotation 181, 187
cross body medial shoulder rotation 183, 187
cross box step-up with dumbbells 178, 187
cross body tricep extension 172, 175, 182, 187
cycling 213

D

Daily Planning Worksheet 116
dead lift off blocks 170-71, 175, 182, 187
degenerative joint changes 95
disc degeneration 109
downswing 14, 104, 140, 209
driving 214, 217

E

econcentric
    loading 193
    strength 196
endurance 135
equipment 96, 119, 134, 153, 176, 191, 204
etiology 91

INDEX

exercises
- bent over row 125, 127-129
- box step-up 162-3, 175
- cross body external shoulder rotation 181, 187
- cross body medial shoulder rotations 183, 187
- cross body step-up with dumbbells 178, 187
- cross body tricep extension 172, 175, 182, 187
- dead lift off blocks 170-71, 175, 182, 187
- Fitter 142-144
- Fitter wobble board 142, 146-148
- 4 point transversus abdominis trainer 107, 116-118
- forward ball roll 121, 127-129
- frontal static lean 124, 127-129
- Greg Johnson's window washer 108, 116-118
- grip trainer 126, 127-129, 145, 146-149
- horse stance alphabet 122, 127-129
- horse stance horizontal 103, 116-118
- horse stance vertical 102, 116-118
- kneeling on Swiss Ball 139, 146-148
- lateral shoulder rotators 169-65, 175, 202-203
- lower abdominal #1-3 111-113, 116-118, 174-75, 186-7, 204, 212-13
- lunge, multi-directional 156-7, 175, 192, 203
- medial shoulder rotation 50, 173-4, 175, 194-5, 203
- medicine ball sit-up pass 193-4, 203
- oblique toss 200-1, 203
- pelvic shift with towel handle 184, 187, 203
- pelvic shift with tornado ball 210, 212
- power ball horizontal chop 207, 212
- power swing 192, 203, 210-11
- prone bridge 123, 127-129
- prone cobra 99-101, 116-118
- prone jack knife 135, 146-148
- prone twister 138, 146-148
- reverse wood chop 168, 175, 198, 203
- rotator cuff Tornado Ball cross body pattern 209, 212
- single arm C.R.A.C. press 180, 187, 204, 211-12
- single arm high cable row & reach 166-7, 175, 179, 187
- soccer toss 208, 212
- standing single arm cable push 154-5, 175
- supine hip extension: feet on ball 120, 127-129
- supine hip extension: knee flexion 133-4, 146-148
- supine lateral ball roll 136, 146-149, 204, 211, 212
- supine single arm Swiss Ball dumbbell press 177, 187
- supine Russian twist 141, 147-149
- Swiss Ball neck training 104-106, 116-118
- Swiss Ball side flexion 185, 187
- Swiss Ball supine hip extension 179, 187
- Swiss Ball reverse hyper-extension 109, 116-118
- Tornado Ball horizontal chop 207, 212
- walking lunge with twist 205-6, 212
- wood chop 160-61, 175, 196-8, 203

external hip rotators (cigarette butt stretch) 62
external rotation 34, 44

F

facilitation 40
fast twitch muscles 74
Feldenkrais 18, 77
female golfers 2, 3, 71, 148, 208
Fitter 127, 136-8, 142-144
Fitter wobble board 142, 146-148
flexibility 5-7, 18, 19, 21, 149, 170, 215-16
Flexibility Tests: table 44
- Apley scratch test 32
- arm raise test 40
- cigarette butt test 38
- McKenzie press-up 42-43
- neck rotation test 29
- neck side flexion test 29
- pectoralis minor and major test 30-33
- side bend test 38

INDEX

spinal rotation test 34
supine knee extension test 36
sweetheart test 30
Thomas test 35
thoracic extension test 41
waiter's bow test 37

follow-through 13, 15, 17, 32-35, 38-43, 71, 87, 104, 140
forearm 126
forward ball roll 121, 127-129
forward head posture 12
4 point transversus abdominis trainer 107, 116-118
frontal plane 87
frontal static lean 124, 127-129

G

gastrocnemius stretch 67
glenohumeral joint 24
golf stability 92
golfer's elbow 151
Greg Johnson's window washer 108, 116-118
grip trainer 119, 126, 127-129, 145, 146-148
groin (Indian sit) 63, 75
groin rocking 64

H

hamstring, supine knee extension 65, 75
hamstring (waiter's bow) 65
hip flexors 58, 71, 121, 135
hip joints 25, 157
hip and pelvis integrator 83
hip rotation 38, 86
90/90 hip stretch 60-61, 75
hook 71
horse stance alphabet 122, 127-129
horse stance horizontal 103, 116-118
horse stance vertical 102, 116-118

I

impingement syndrome 43
integration 96
internal hip rotators 63
isolation 96

J

joint capsule 22
joint soreness 191

K

knee pain 215
kneeling on Swiss Ball 139-40, 146-148
kyphotic posture 94

L

lateral rotation 22
lateral shoulder rotators 52, 71, 164
latissimus dorsi 40, 44, 55, 75
law of facilitation 40
leg flicks 89
length/tension relationship 36
levator scapulae 30, 48, 75
limited spinal rotation 38
low back pain 84, 91, 215
lower abdominal exercises 110-14, 174, 186, 212
lumbar disc bulge 59
lumbar erector stretch 59, 75, 101
lumbar spine 44, 171
lunge, multi-directional 156-4, 175, 192, 203
lunge stretch 58

M

McKenzie press-up 42, 68
medial rotation 22

INDEX

medial shoulder rotation 50, 75, 173-4, 183, 194-5
medicine ball exercises
    bouncing wood chop 196-7
    lateral shoulder rotators 202
    medial shoulder rotators 194
    oblique toss 200
    power swing 192, 210, 212
    reverse wood chop 198
    sit-up pass 193-4, 203
    soccer toss 208
motor program development 5, 6, 73
motor patterns 160
muscle balance 18, 28
muscle energy exercises 77, 78, 84
    foot/ankle warm-up 88
    golfer's neck/trunk trainer 78, 91, 99-100
    hip/pelvis integrator 83
    hip rotations 86
    leg/arm flicks 89
    shoulder clock 85
    shoulder/spine integrator 80-82
    weight shifting 86-87
muscle imbalance 35

N

neck extensors 49
neck flexors 193-4
neck rotators 46
neck rotation test 29
neck side flexors 47
neck side flexion test 29
neck trunk trainer 77, 84, 90
nervous system 9, 25, 75, 76, 107, 170-71, 210
neuromuscular 96, 98, 115, 119
Newton's law 168
90/90 hip stretch 57-8
non-physiological joint motion 88

O

obliques 57, 185
open chained exercise 5, 7
optimal visual focus 16
overtraining 153

P

pectoralis major 53
pectoralis minor 50, 53
pectoralis minor and major test 33
pelvic girdle 83, 141
pelvic shift with towel handle 184, 187, 199, 203
pelvic shift with Tornado Ball 199-210, 212
phasic muscles 27
Physio-toner 98
pitching 217
planes of movement 21
plumb line 94
posture 8-18, 93, 94, 99, 105, 130, 135
    address 11-12, 21, 36, 71
    dynamic 9, 13, 18, 139, 170
    endurance 120, 125
    good 5, 8, 10, 94, 96
    kyphotic-lordotic 94
    muscles 101
    static 9, 10, 123
    sway 93, 95, 96
power 18, 19, 38, 41, 83, 130, 151, 189, 190-212, 213, 215-17
power swing 192, 203, 210-11
Power web 119, 126
prone bridge 123, 127-129
prone cobra 99-101, 116-118
prone jack knife 130, 146-148
prone twister 138, 146-148
prevention (of injury) 92
prone cobra 101
proprioception 104, 143

INDEX

protracted shoulders 12
pull 71
pushing 71
putting 65, 214, 217

Q

quadratus lumborum 185
quadriceps, supine 66
quadriceps, Swiss Ball 66-67

R

range of motion 78
rebounder 193-95, 200-204
rectus abominis stretch 56
relative timing 199
reverse wood chop 168, 175, 198, 203
rhomboids 52
righting reflex 136-37
rotation 24
rotator cuff muscles 164, 209
rotator cuff Tornado Ball cross body pattern 209, 212
rules of prevention and performance 92

S

sacroiliac joints 83, 87
S.A.I.D. principle (Specific Adaptation to Imposed Demands) 190
scapulothoracic joint 24
senior golfers 71, 97
shank 71
shoulder 164, 167
    extensors 121
    girdle 108, 125, 135, 136, 151
    lateral shoulder rotators 164-5, 175, 202-203
    medial shoulder rotation 173-4, 175, 194-5, 203
    pain 215
    range of motion 24

shoulder clocks 85
shoulder spine integrator 80
side bend test 38
single arm C.R.A.C. press 180, 182, 204, 211-12
single arm high cable row & reach 166-7, 170, 175, 187
slice 71
slow twitch muscles 74
soccer toss 208, 212
soleus stretch 67
specificity 190
speed 4, 216
spine 78, 121
spinal range of motion 24, 35, 38, 58, 86, 202
spinal rotation test 34
stability 18-19, 92
    dynamic 92, 96, 97, 119, 130-1, 189-90, 210-211, 213-17
    static 92, 96, 97, 130-1, 189-91, 213-17
    postural 93, 110, 149
stabilizer muscles 6, 98, 103, 107, 120, 162, 189
standing hamstring stretch 65
standing single arm cable push 154-5, 175
strength 18, 19, 104, 123, 130, 136, 149-53, 213, 215-17
stretches: table 71
    calves 67
    external hip rotators (cigarette butt stretch) 62
    groin (Indian sit) 63, 71
    groin rocking 64
    hamstring, supine knee extension 65, 71
    hamstring (waiter's bow) 65
    90/90 hip stretch 60, 71
    hip flexors, lunge 58, 71
    internal hip rotators 63
    lateral shoulder rotators 52, 71
    latissimus dorsi 55, 71
    levator scapulae 48, 71
    lumbar erectors 59, 71
    McKenzie press-up 68
    medial shoulder roators 50-51, 71

INDEX

neck extensors 49
neck rotators 46
neck side flexors 47
oblique abdominals 57
pectoralis major 53
pectoralis minor 50
quadriceps, supine 66
quadriceps, Swiss Ball 66-67
rectus abdominis 56
rhomboids 52
thoracic spine, foam roller mobilizations 69-71
trunk rotation 57, 58, 75
wrist extensors 54
wrist flexors 45-76
stretching 45-76
    developmental 26, 72, 76
    maintenance 26, 72, 75
    muscle energy mobilization 26
    post-event 26, 75, 75
    pre-event 26, 73, 75
subscapularis 173
supine hip extension: feet on ball 120, 127-129
supine hip extension: knee flexion 133-4, 146-148
supine knee extension test 36
supine lateral ball roll 136, 146-148, 211, 212
supine single arm Swiss Ball dumbbell press 177, 187
supine Russian twist 141, 146-148
supine quadriceps stretch 66-67
sweetheart test 30
sweet spot 4, 216
swing
    arc 38, 41
    faults 22, 33, 38, 97, 151, 216
    mechanics 48, 200
    path 4, 216
    plane 21, 29, 38, 39, 41, 189, 190
    timing 21
swing progression 89

Swiss Ball 51-3, 56-7, 66-67, 104-106, 108-10, 116-18, 119-20, 123-24, 127-30, 177, 185, 187, 193-4, 203
    sizing 132
Swiss Ball neck training 104-106, 116-118
Swiss Ball side flexion 185, 187
Swiss Ball supine hip extension 179, 187
Swiss Ball reverse hyper-extension 109, 116-118
Swiss Ball quadriceps stretch 66-67

T

taping (lumbar spine) 61, 171
Thomas test 35, 57
thoracic extension test 41
thoracic kyphosis 12
thoracic spine, foam roller mobilizations 69-71
tilting reflexes 139
tonic muscles 27, 46, 98
Tornado Ball 207, 209, 210, 212
Total Gym 152
transversus abdominis 107, 158, 186
Tri-axial relationship 15
trunk rotation 57, 160
    stretch 57-58, 75
    test 79

W

walking lunge with twist 205-6, 212
waiter's bow test 37, 65
warm-up 85-90
    pre-golf 77, 85, 90
    dynamic 77
    swinging progressions 90
weight shifting 86
whole in one concept 5
wobble board 142
wood chop 160-1, 175, 196-98, 203
wrist extensors 54
wrist flexors 54